'Mary "rights-based" Robinson has been right about a lot of things . . . human wrongs and human rights . . . not just in theory but in practice and in action, whether for Ireland as President, for the UN as Rights Commissioner, at GAVI as Chair, or at the Elders. We need more Marys. Hail Mary.'

Bono

'Mary Robinson has all the Gaelic attributes; she is warm, friendly and engaging. She is really passionate about human rights, especially the rights of women and young girls, and cares deeply about the environment. This is her story, the story of the lawyer who rose to be Ireland's first woman President and the UN High Commissioner for Human Rights.'

Archbishop Emeritus Desmond Tutu

'Like many English words, the word "bold" – brave to the British – acquired new depths across the Irish Channel. It gained a sense of rebellion and a twinkle of mischief. Mary Robinson is a bold woman. Whether taking on the oppressive orthodoxy of Catholic dogma embedded in the Constitution, the rights of women, the oppressed and the dispossessed, she has been bold, courageous, compassionate and fiercely determined to do what is right. Even when she was a young girl playing the part of Batman, she took on the entire forces of evil and has been a fearless champion of those in need ever since. This is an extraordinary story of an extraordinary woman, who is also one of my heroes.'

Peter Gabriel

'*Everybody Matters* has all the hallmarks of the woman herself. Concerned for the truth, unflinching in the telling of it and passionate to achieve a fairer world. I hope it inspires as many as possible.'

Sir Richard Branson

D0320124

Mary Robinson served as the seventh, and first woman, President of Ireland from 1990-1997, and the United Nations High Commissioner for Human Rights from 1997-2002. Robinson has been Honorary President of Oxfam International since 2002, and has chaired numerous bodies including the GAVI Alliance, vaccinating children worldwide, the Council of Women World Leaders (of which she was a co-founder), the International Institute for Environment and Development, and the Institute for Human Rights and Business. A former President of the International Commission of Jurists, Robinson serves on the board of the Mo Ibrahim Foundation, which supports good governance in Africa, and is a member of the Elders, an independent group of global leaders brought together by Nelson Mandela. A member of the Royal Irish Academy and the American Philosophical Society, she is the recipient of numerous awards and honours, including the US Presidential Medal of Freedom, and the Indira Ghandi and Sydney Peace Prizes, and has been Chancellor of Dublin University since 1998. In 2013 Mary was made Special Envoy for the Great Lakes region of Africa by the United Nations Secretary General Ban Ki-Moon. She is married to Nick Robinson with three children and four grandchildren. Also President of the Mary Robinson Foundation – Climate Justice, she lives in Dublin and Mayo.

Tessa Robinson is a graduate of Trinity College Dublin and the King's Inns. She practised as a barrister for ten years before becoming a freelance writer and copy editor. She is Mary's daughter and lives in Dublin with her husband and two children.

Everybody Matters

A Memoir

MARY ROBINSON

WITH

TESSA ROBINSON

HODDER

First published in Great Britain in 2012 by Hodder & Stoughton
An Hachette UK company

First published in paperback in 2013

1

Copyright © Mary Robinson 2012

The right of Mary Robinson to be identified as the Author of the Work has been
asserted by her in accordance with the Copyright, Designs and Patents Act 1988.

A CIP catalogue record for this title is available from the British Library

ISBN 978 1 444 72333 5

Printed and bound by Clays Ltd, St Ives plc

Hodder & Stoughton policy is to use papers that are natural, renewable
and recyclable products and made from wood grown in sustainable
forests. The logging and manufacturing processes are expected to
conform to the environmental regulations of the country of origin.

Hodder & Stoughton Ltd
338 Euston Road
London NW1 3BH

www.hodder.co.uk

To Rory, Amy, Otto, and Kira

Contents

1

Beginnings

Where, after all, do universal rights begin? In small places, close to home − so close and so small that they cannot be seen on any maps of the world. Yet they are the world of the individual person; the neighbourhood he lives in; the school or college he attends; the factory, farm or office where he works. Such are the places where every man, woman, and child seeks equal justice, equal opportunity, equal dignity without discrimination. Unless these rights have meaning there, they have little meaning anywhere. Without concerted citizen action to uphold them close to home, we shall look in vain for progress in the larger world.
Eleanor Roosevelt, at the United Nations, New York, 27 March 1958

I WAS born and spent my early years in the west of Ireland in just such a place: Ballina, then a town of some six thousand inhabitants, in north County Mayo − a small town in a small country on the western periphery of Europe. It had its strong local history and legend, and was, naturally, the centre of the universe for me as a young child. When, later, I first read those famous words of Eleanor Roosevelt, in my boarding school library, I felt a frisson of excitement as I identified Ballina with the small places she spoke of, put side by side with the grand ideals of the Universal Declaration. Looking out the library window, I saw her words as a window into a new world, where concerted citizen action would ensure that everybody mattered. We are all shaped by our early influences, many of which we absorb without being conscious of or fully understanding them.

Mine was a privileged family living in a community that had its share of grinding poverty but also a cohesion stemming from people's faith and involvement in the life of the parish. My parents were both doctors. Aubrey de Vere Bourke, my father, was raised in a house called Amana, the home of his parents, Henry Charles Bourke and

Eleanor Bourke née Macaulay. 'HC' and 'Nellie' to their friends, they were known by their children as the Pater and the Mater. Amana looked down on the Ridge Pool, celebrated as a stretch of the River Moy that had made Ballina famous for its salmon fishing.

Like his six siblings, my father had been educated in England. He had boarded at the Jesuit school, Mount St Mary's, and had taken his medical degree at Edinburgh University. As a young doctor he worked in London before taking up an internship in the Coombe Hospital in Dublin. The Coombe catered for the very poorest in Dublin, people living in tenements of the kind found in Sean O'Casey plays. My father often talked about the intense poverty: rats in the homes of patients he visited, the overcrowding. At the Coombe Hospital he met another doctor, Tessa O'Donnell.

Tessa O'Donnell, my mother, was from Carndonagh, County Donegal on the Inishowen Peninsula at the northernmost tip of Ireland. Her parents, Hubert and Winifred O'Donnell, were shop-keepers and early supporters of the credit union movement. A family of high achievers, they succeeded in putting five of their children through university, four of them becoming doctors.

My mother studied her medicine at University College Dublin (UCD). She liked to joke that it had taken her so long to qualify because she had been in no hurry, enjoying the social scene, captain-ing the UCD hockey and tennis teams, and preparing the teas for the rugby club. I am not sure how many years were involved, but certainly she knew a great many people, mainly living in Dublin, whom we would meet on visits there. Before finding a position at the Coombe she had worked as a locum doctor on one of the islands off the coast of Donegal, Aranmore, where she served a community living in terrible poverty that lacked any other kind of medical support.

By all accounts, Tessa O'Donnell fell head over heels in love with Aubrey Bourke and was the moving party in their relationship. Being some years older than he she drew him out of himself. Aubrey was born in the year 1914. Tessa was not forthcoming about her age. She used to tell us that she was born the year the *Titanic* sank, 1912, but we now know that she was born in 1908. Aubrey was handsome and athletic but quite reserved; he had trained for the priesthood with the Jesuits – he never spoke of why it did not work out – before turning to study medicine. Where Aubrey was serious and reserved, Tessa was extroverted and fun.

They married in Dublin on 18 January 1940. Bishop Naughton, who may also have married HC Bourke and Nellie Macaulay a generation earlier, officiated at their wedding, which took place at University Church on St Stephen's Green. My father's brother, Roddy, was the best man, and my mother's sister, Florrie, was the maid of honour. Although my father later made it out to be a small wedding – these were austere times early in World War II – it was, according to my uncle Roddy, a big affair, with a wedding breakfast in the Shelbourne Hotel and photographs in the newspapers.

When the newly weds returned to Mayo, my grandfather HC felt that my father's best prospects would be to emigrate (as his siblings had done or would do). My father's eldest brother, Paget, was already doing well in the colonial service and would later become a chief justice and be knighted by Queen Elizabeth II. Another brother, Hal, was a doctor in England. Roddy would emigrate to Australia, and the youngest, Dennis, would end up in Brazil. Even the two sisters, Ivy and Dorothy, who became nuns, went to India and England respectively. However, my father was adamant that having been educated outside Ireland, of which he was appreciative, he was going to remain in the town of Ballina and practise medicine from there, as his maternal grandfather, Roger Macaulay, had done.

So Aubrey and Tessa acquired No 2 Victoria Terrace, a semi-detached house on the quay in Ballina, facing St Muredach's Cathedral, across the River Moy. My father fixed a brass plaque to the door, opened for business as a general practitioner, and my parents set about starting a family.

I was born in May 1944, the third child and only daughter. My two older brothers were Oliver and Aubrey, and my two younger, Henry and Adrian. The five of us were born within six and a half years.

Victoria House was built as one property in or about 1840, shortly after the accession of Queen Victoria to the throne; hence the name, which has somehow survived over the years. That half of the property my parents initially acquired – No 2 Victoria Terrace, – was, by coincidence, the house my grandfather HC had lived in and married from before acquiring Amana.

On entering through the front door of No 2 (painted red), you came into a hall with a tiled floor. On the immediate right was the dining room, behind that a darkish waiting room with windows looking out onto the neighbouring premises of Reilly and Boland Timber Merchants. Behind that again was the kitchen, scullery and a back door out to a yard with a large shed for storing turf and household bits and pieces. Up the stairs, at the turn, a flight of four or five steps led to my father's surgery. Patients were let into the house either by my mother or by the maid, and their names were kept in a big book just at the door of the kitchen so that as each patient went out, my father would stride in there, read the next name in the book, and call it out in the waiting room.

Upstairs was a fine drawing room with a mahogany and oak floor, pale damask curtains, a piano, an open fire, and good furniture, and in this room my father would read to us from Rudyard Kipling and other children's classics. We would do our musical lessons there, and were allowed to have sweets from the tin box on the piano after lunch on Sunday, and only then.

On the next floor up were our bedrooms. Henry and Adrian shared one, Ollie and Aubrey another, and I had my own small, narrow bedroom with a window looking out front, over the River Moy, to the cathedral across the way. Up the next floor again was the attic level, with a bedroom for the maid and general storage for old clothes, toys and so on, a good place for children playing mystery or adventure games.

My parents bought No 1 Victoria Terrace in 1960 when our neighbour died, and Mummy oversaw the work to effectively rejoin the two houses.

I remember feeling as a child that my father worked constantly. He would receive patients from early morning, often interrupting his breakfast; his lunch, too, was frequently interrupted, and he would take urgent calls in the middle of the night. There were times when he rode out on horseback, if the conditions were not amenable to driving, as he had done during the war when petrol was in short supply. We were trained from an early age to be quiet and respectful around his patients. Thanks, I suspect, to my mother's keen management, he became doctor, and good friend, to many of the wealthier families in and around Ballina, as well as to the ordinary middle- and working-class citizens. Despite his reserve, he had a good manner with his patients and built up a strong professional reputation. He was a skilful diagnostician, read seriously the current medical journals, and was excited by medical developments. He would tell with wonder, for example, of his early encounters with the new medical miracle of penicillin.

My father also had good business sense. When a number of chemists in Ballina approached him with a proposal for supplying pharmaceutical drugs to the West of Ireland, he took on the role of chair of the venture, which they called United Drug. My father remained chairman for twenty years, and United Drug became a successful company, operating internationally.

When he wasn't working, my father would be tending to his garden, an open plot of land beside our house and across a laneway. There he grew flowers – roses, sweet pea, tulips, and lilies – and had a thriving vegetable plot, with a greenhouse where he grew carnations, tomatoes, and cucumbers. My mother would love to sit on the steps in the garden, smoking cigarettes, chatting and watching him at his work. One of my chores as a small girl was to help my mother arrange the flowers in the house.

My father's other passion was for horses. He was athletic and enjoyed playing and watching sports of all kinds, but he derived the most joy from horses. He took up riding relatively late, into his thirties, I believe, but was an excellent horseman and won jumping competitions at horse shows in Ballina and other towns in the region. On the day of a big race such as the Grand National, he would close the curtains in the small sitting room we called the 'smoke room' (in imitation of the one in my grandfather's house), put the race on the television, and roar at the black-and-white screen as if he were there in the stands. Unlike my mother, he rarely left

Ballina in those days, as he was working, but occasionally he would get to some race meeting or other.

While he did ease up considerably in his later years, he never ceased to be a doctor, abiding by his own adage that it was better to wear out than to rust out. Not until his eighty-eighth year, some two weeks before he died in Castlebar Hospital surrounded by children and grandchildren, did he write to his few remaining private patients informing them of his retirement from practice.

My mother was not a particularly handsome woman. Her attractiveness lay in her charisma; she had an outgoing, warm personality and a great sense of humour. She was one of those people in whom others easily confided because she knew how to listen and how to ask questions without appearing inquisitive.

She carried with her at all times a fat brown leather handbag, and if you got it open you'd find Kingsway cigarettes, a range of powders and lipsticks and the like, and a big wad of money, which she used to pay for everything. She possessed great organisational skills, reputedly furnishing her new house in Victoria Terrace from one shop in one afternoon. She ran our home as a tight ship. She would buy her household goods in bulk and store them in the sheds at the back of the house. Every month, a big load of turf for the fire would be delivered, which my brothers would help stack in the shed. We five children were the centre of her world, and she and my father worked hard with the single aim of ensuring that we had the best of everything. And so, I suppose, we did, attending private schools and university, always well turned out and looked after.

In addition to managing my father's surgery practice, my mother helped my grandmother run a number of charities in the town and provided the flowers for the cathedral on almost a daily basis. Years later, after my grandmother died, my mother took over the running of my grandfather's house in Amana; they were great friends. My mother was fulfilled by all this, and did not mind giving up the practice of medicine. She used to joke that she preferred to practise on us! In any case, it would have been unusual at that time for a married woman to continue in practice; in those days, for example, women who married were still required to resign from their jobs in the civil service.

My mother loved to shop for me, though I was not a good subject; I was a book worm and a dreamer, abstracted and not interested in clothes. She recognised and accepted this, but rather than be disappointed that I did not share her interests, she took on the role of supporting and organising these practical aspects of my life – well into my adult life – ensuring that I was well dressed. She would drive me up to Dublin to take me shopping, and I remember many walks up Grafton Street that seemed to take forever. After a few yards we would meet her first acquaintance; they would embrace warmly and begin to gossip. I would hang back and either open my book to read or stand there daydreaming. This might happen a few times until eventually we reached our destination, usually the department store, Brown Thomas. There she would greet the people behind the counters as old friends with whom she had shopped over the years. She never simply said, hello, Mrs so-and-so; she would inquire about her daughter who wasn't so well a year ago, and was she still worried about such-and-such. I recognise now that my mother was a born politician, able to relate to partial strangers and have them remember her as she remembered their names and the details of their lives.

As we grew older, my brothers and I increasingly appreciated that our mother had a wicked sense of humour and was much more likely to tolerate our being naughty than was my father. She adored my father and had him on a pedestal, as we all did. But he was more straight-laced, and less inclined to see any humour in vulgarity or adolescent jokes.

My mother enjoyed dancing and liked to stay on at parties until the end, even though my father was keen to go home. She also took a huge interest in our friends. I remember when a close friend from boarding school, Mary Courtney, came to stay with us, she was much taken by my mother, how loving and how interested she was in everything that Mary and I were doing, and they formed a very strong attachment.

In the kitchen, my mother loved to recount the experiences of her day up shopping in the town or over in Amana with my grandfather. She would stand in front of the AGA cooker, lifting her skirt so that her bottom warmed nicely. When any of us came in, she would put her arms out and give us a big hug. That was the warmest place in my early memory: my mother's arms around me, her backside to the fire, and the world secure.

Despite loving her deeply, the problem I had with my mother, as I grew older, lay in our differences of opinion about family and status. Her father had died the year I was born, but I remember well 'Granny in the North', as we used to call her mother, a neat, handsome woman who always wore black but who looked well and content in herself. One of my excitements on a visit to her was being allowed to serve in her shop, selling sweets, making accurate calculations of what things cost, and putting the money in the till. In Ballina, my mother was friendly with many of the ordinary folk and tradespeople of the town. But she seemed to seek out the wealthier families, often Protestant, aristocratic people living in the vicinity of Ballina: Con Aldridge of Mount Falcon and her husband, the Major; the Armitages; the Trittons; the Pery-Knox-Gores; Jack Philipson Stowe, a baronet. My mother insisted that the Bourkes, whose ancestors had been part of the twelfth-century Norman invasion of Ireland, and claimed descent from Emperor Charlemagne, were more significant than other families and she tried to instil that in us. The more she did this, the more something in me resisted it. I did not fight my mother about it; I gave up arguing with her, but deep inside I queried her sense that because the Bourkes had a crest, because the local Protestant church contained memorial tablets of Victorian generals – one, my father's great uncle Paget, had been captain of the Queen's bodyguard at Buckingham Palace – that somehow we were special.

My mother had explained to us that my grandfather and his siblings were the first Catholics in the Bourke line, William Orme Bourke having married a devout Catholic who had ensured that all her children were baptised Catholic. At school, I learned about Ireland's struggles for freedom, about Michael Davitt and the Land League, about the 1916 rebellion. These events interested me more than the idea that somehow our genes were more important or of greater quality than others'. I took comfort from the fact that on my grandmother Macaulay's side, her uncle Thomas Macauley had been jailed as a Fenian, for belonging to the Irish Republican Brotherhood, an organisation committed to overthrowing British rule in Ireland. (Officialdom's mis-spelling of his name was perhaps appreciated by my great-grandfather Roger, his brother.)

I had difficulty at the time putting my finger on exactly what made me so uncomfortable about the sense of status that was instilled in us. We lived a privileged life, no doubt, and enjoyed the benefits of it. But I questioned the seeming unfairness of the chance of birth. It occurred to me that a sense of family status didn't quite tally with the teachings of Christianity. According to Christ, people were all equal and should be treated equally but from early on I could see that this was not the case and that, indeed, there seemed to be a great deal of unfairness in the world.

Growing up, I identified more with my father – and of course I admired him greatly. I felt, in my seriousness and reserve, in my application to intellectual work, and in my interest in the welfare of people, that I was following in his footsteps. However, much later, when I realised, as a candidate for the Irish presidency, that I would have to open up and show more of the inner self – that shy inner self that I had not made available to people – I followed in the footsteps of my mother. I reflected on her openness to other people and on how much she gave of herself. Once I started opening up like that I never went back.

The cine-films that my father loved to take when we were on holidays in Donegal, or at the beach in Enniscrone, showed that my brothers tanned a beautiful brown and were blue-eyed and had wide, engaging smiles, whereas I was freckled and plump, and went red when exposed to the sun. I knew that my brothers were better looking than I, and I felt this all the more because I was the only girl. Nonetheless, I discovered that I was quick to learn. I learned to read before my clever but laid-back older brother, Aubrey, something I childishly delighted in showing off about.

As a family of five children, we generally split into two groups: my two older brothers as one group, and my younger brothers and me as the other. Naturally, among these brothers, I was a tomboy. They seemed to be content just to play, whereas, from a very early stage, when I was kicking a football or running along the beach, I was also thinking about the world or the discussions my parents had at table the night before. I was always testing myself physically, trying to go farther, jump higher, run faster. And yet the real testing I was doing was to figure out how I could be better: how I could somehow compensate for the fact that I was the freckled, plump one who needed to be able to prove she could be loved. Yet I had no reason to doubt that I was loved, because my parents were extraordinarily affectionate to all five of us, and as my mother would often say I was the favoured, only girl.

Religion played a central role in our family life. We attended Mass, in Latin, every Sunday. In the evenings we would kneel in our own set places in the drawing room, one parent on each side of the fire-place, and say a decade of the rosary. We said grace before and after every meal. Crucifixes hung in most of the rooms, as well as pictures of the Sacred Heart and the Virgin Mary. During Lent, these would be covered with black cloth, and my parents would wear mourning clothes and fast. My mother took great pride in providing the flowers for the cathedral altar, particularly on Sundays. My grandmother and grandfather were daily communicants. I still remember going to the first midnight Mass held in Ballina, when I was about eight years old. I was so proud of the fact that my grandmother was willing to take me and have me sit with her and my grandfather.

We grew up with this sense of the importance of the Church, the importance of prayer and faith. We did not question it. This was Ireland in the 1950s. Our relatively new Constitution – (enshrined in 1937) – placed God and the Roman Catholic faith at its centre. This was a stratified society where people knew their place and did not question it. Priests held a status that was high and implicit and trusted. People did not look behind the white collar. In this Ireland, single mothers were pariahs who had their children taken from them to be brought up by good Catholic parents while the mothers paid for their sins, working as slaves to the religious orders in institutions such as the Magdalene laundries. Of course, we have since found out how some priests abused this power, preying on the vulnerable behind closed doors, and how some were protected from the force of the law by those in the highest echelons in the Church. This is an episode of deep shame in the nation's history (and is more shameful, in my view, than the recent humbling intervention in Ireland's economic crisis by the International Monetary Fund, the European Central Bank, and the European Union).

My brothers and I attended Miss Ruddy's primary school in Ballina, a private school for middle-class children, until the age of eleven or twelve. It being a small town, we all knew the families of our classmates. Particular friends, who remained so long afterwards, were Paddy Murphy and Heather and Robert Morrow. Paddy was a warm-hearted, generous boy, better off than most of us, as his family owned Murphy's Flour Mills. Having more pocket money, he would buy a large bag of sweets when we went to matinees in the Savoy or Astoria cinema, and share them along the row of expectant kids. Later he was to be a generous, much-loved citizen of Ballina, until his untimely death from cancer. Heather and Robert Morrow were children of Willie and Hope Morrow, close friends of my parents. We spent many happy times playing in the grounds of their farm-house down the Sligo Road. I gradually appreciated that they were Protestant, but this was not an issue at school, except that they did not join in the daily singing of the Angelus.

The prevailing attitude at Miss Ruddy's in those days was that we were somehow 'better' than the majority of children in Ballina who were going to the convent school. When we met children from the convent, both sides would call names and exchange insults. Even though the divide was deeply instilled in us, part of life in a small town, the unfairness of the better-off jeering the less-well-off, made me uncomfortable.

Miss Ruddy's was at first a one-teacher school, until Miss Ruddy took on an assistant, Miss Conmy. We learned everything by rote – times tables, poetry, the Ten Commandments, the entire Catechism – until we knew them, and if we failed to learn them, we were at fault. Miss Ruddy was small and fierce, and had a cane, which she used liberally. The girls wore skirts, and the boys, short trousers, and we would be smacked on the back of the legs. At times my mother would bring a box of chocolates to Miss Ruddy to sweeten her and try to reduce the caning of her children, but to my mother's dismay, my brother Aubrey still managed to be irrepressibly spirited enough to warrant quite a lot of caning.

Our father, too, was a strict disciplinarian. He had commandeered a toy rubber knife, long and flexible, which he used frequently to smack our hands when we were naughty, and occasionally our bottoms. I recall the dire warning we regularly exchanged: 'Don't do that or Dad will take the knife to you!' That was a normal way in which parents coped with children at the time. It was an accepted part of life; it was unpleasant, but there was no questioning it.

We learned, and were taught by example all around us, to work hard. But we were children, so naturally we developed ingenious schemes to get out of work, or to conspire to get our own way. One of our chores was to pick the gooseberries in HC's garden. To avoid getting our hands and arms shredded by the thorns, one of us would crawl under the bush and give it a mighty whack with a stick so that every single gooseberry fell to the ground – ripe or not. We got into trouble when caught at this one morning by my father. We also had a taste for the horror B-movies playing at the Astoria cinema, which were strictly out of bounds. Adrian, as the youngest, would be sent to my father to ask for the money, pretending we wanted to go to a Roy Rogers, *Boy's Own*-type picture, and then we would head straight for the film we really wanted to see. On one such occasion we were engrossed in a vile, *Frankenstein*-themed horror film. About halfway through, all the voices ground to a halt, the film stopped, and the lights went on. The proprietor, Mr Mulligan, then called out, 'Would Mrs Doctor Bourke's children please stand up and leave the cinema.' We looked around to see our mother standing at the back with Mulligan. In utter shame, we stood up, red as beetroot, and slunk out of the cinema, to reprimands and liberal smackings.

One further incident has stuck in my mind. After lunch one day, my younger brothers, Henry and Adrian, and I were playing too loudly and complaining of being bored in the house. My father wanted us out from under his feet. He barked impatiently at me, 'Go on out and take your brothers for a walk.' His reproof stung, and I decided I would show him. I would walk my brothers about eight miles away, to Enniscrone, a coastal town in County Sligo. We were roughly eight, six, and five at the time. Off we three marched, across the bridge and down the quays and onto the road to Enniscrone. Becoming quite thirsty along the route, we called in to a house and, explaining that we were Dr Bourke's children, asked for some milk, which we were courteously given. On and on we walked until we did, eventually, reach Enniscrone. It took us several hours. I remember putting sand from the beach into my pockets to prove that we had made it all the way. Then the realisation came that we would somehow have to get back. There was nothing for it but to call home. After begging for a threepenny bit for the phone box from the house nearest it, we crowded into the phone box together. I dialled the operator and asked to be connected to Dr Bourke's. My father answered, and we had a very short conversation indeed. 'Where are you?' 'Enniscrone.' 'What are you doing there?' 'You told us to go for a long walk.' 'Stay there.' We stayed where we were, and my father arrived, turned the car, and said, 'Get in.' The three of us got in to the car, and he never said a word. We weren't punished that evening. I think he was secretly so proud of us he didn't quite know what to say or do.

We were not always so wilful, but we were a lively bunch by nature. The dining room was our playroom when we were young children. All our toys were kept in a cupboard, and we would take them out and take them to our corner of the dining room table. I was an imaginative child and not only played with dolls' houses and the toy cars and trains of my brothers but also invented games that Henry and Adrian would join in on. One of our favourites was Batman and Robin. Henry would be Robin, I would be Batman, and poor Adrian was the villain, usually without his full consent. When we played Roy Rogers, I would be Roy, Henry would be Tonto, and Adrian would again be the baddie. I was not particularly conscious that we were playing boys' games; I enjoyed games with a bit of adventure to them, and I enjoyed competing with my brothers. One game involved our having to move all around the room without our feet touching the ground, so that we would leap to and from pieces of furniture. With basic wooden toys and no television, we played these games from our imagination, and dressed up and staged plays for our parents.

Playing in the dining room, I liked to listen as my father saw patients out from his surgery. I noticed from a young age that, during these times, he changed his speaking voice. His natural speaking voice was polished. But when he was letting out patients, I would hear a slower, more West of Ireland voice. It seemed to me he was making a particular effort to communicate with people. I was learning, too, from what my father said to my mother over the dinner table, that some diseases were, for the patients, a matter of shame: families would not talk about the fact that somebody was very ill with cancer. I listened and learned about various illnesses, such as tuberculosis, which patients talked about in whispers. This exasperated my father, as did the frequent question to him when he delivered a baby: 'Doctor, is it a boy or a child?' I think he talked about it to my mother to get it off his chest. She was a good listener, empathising with his frustrations. She showed him great affection, always putting her arms around him. And he would be mollified by her and would accept that that was the way people felt: ashamed to have illnesses like cancer or tuberculosis, prejudiced in their preference for a boy-child. Of course, like any good husband and wife, my parents were capable of a lively quarrel from time to time, and when this was happening we would duck out of the way and wait until the atmosphere had cleared.

From about the age of seven, I loved going out on calls with my father, and observed how long he spent in people's homes. These were often run-down houses with outside lavatories, and some of the patients were extremely poor. Yet my father would stand there at the door and converse for a long time, perhaps with the woman of the house, after he had been to see whoever was inside. Sitting in the car, I was impatient for my father to return, but I also realised that he cared deeply about each of his patients. Later I learned that he would charge his wealthy patients a good fee, so as to subsidise those who could afford little or nothing. This explained the appearance of cabbages, chickens, and the odd bottle of poteen given to him close to Christmas.

At that time, the 1950s, it was not difficult for professional families to find maids from country families who were keen to work. My father needed someone to open the door to patients and take their names and see them in to the waiting room. The role was not really that of a professional receptionist but was one of the duties the housekeeper/ maid would carry out. We did have other maids when I was younger, but the one that mattered, for the rest of her life and a good deal of mine, was Annie Coyne. I met her mother and father on calls out to Knockmore. The Coynes were salt-of-the-earth people; decent, honest, and hardworking. We called Annie Coyne 'Nanny' when we were growing up, and subsequently she became nanny to my own children. Nanny's mother, old Mrs. Coyne, was quite a personality, and we used to enjoy visiting her. Showing us her warm smile, with her front teeth missing, she would always produce a piece of cake (usually currant cake), which made her very popular.

Nanny was religious, a frequent, if not daily, communicant. She gave most of her modest wages to the Church in one way or another. She herself did not marry. She wrote with some difficulty, and yet she read most of the novels and serious biographies that my brothers and I were reading as teenagers and later, when we went up to college in Dublin. Nanny was a rock of common sense and country-style wisdom. She rarely chastised us as children, and indeed she spoiled us in many ways by how well she looked after us. But she was strict with us, too, and would insist, for example, that it was good for us to help her clear the table and do the washing up. She was a wonderful cook; no one could bake a sponge cake like Nanny. She was loyal, hardworking, and showed us a gruff kind of affection that we knew was deeply felt.

As she was only about ten years older than I, Nanny played the role, to some extent, of big sister. So that when I had my first period, at the age of about twelve, playing tennis with my brothers, it was to

Nanny that I turned. She brought me into the bathroom and helped me put on a rather old-fashioned sanitary towel. Although I had some idea of what was happening to me (from a talk with my mother I had not bothered to listen to properly), I was crying from anger and embarrassment. Nanny talked to me with her firm common sense, explaining that this happened to all women and that I was lucky, because if I did not have this I would not be able to have children. At that stage, of course, I was nowhere near thinking about children and, like many teenagers, considered this new development an unwelcome intrusion in my life.

My aunt Ivy was another significant influence. Ivy was the older of my father's two sisters and had become a Sacred Heart nun in Roehampton, in England. A forceful personality with a lively wit and great energy, she had a thin, wiry physique and regularly practised yoga. Well read and a wily card player, she was, like many in the family, – excluding me, – passionate about horses.

I never actually met Ivy when I was a child. It was not until many years later, when I was in my twenties, that she first came home for a holiday and we met each other in the flesh. From the beginning, Ivy had elected to go on the missions, which in those days meant leaving Ireland and effectively saying goodbye forever. You gave yourself to the order for life. She was sent to teach at the Sacred Heart convents in India, first to Bangalore and then to Bombay. At that time these were the schools of the elite. Ivy would send long letters to my father, with whom she was very close, which he read to me, and in my early teens I began my own correspondence with her. She described how the children attending the convent were bright and eager to learn, but it troubled her that they lived very separate lives from the poorer children, many of whom were beggars; how some of the girls had these children carry their books to school for them balanced on their heads.

Vatican II, the ecumenical council opened by Pope John XXIII in 1962 and aimed at updating the Catholic Church and adapting it to modern times, brought about a fundamental change in the role of religious orders. The Sacred Heart order took this particularly seriously, shifting its emphasis to working for the poorest children. Ivy described how much more committed she felt in supporting and educating these children. She would ask my father for help with basic necessities. My father would buy a large boxful of bars of green Palmolive soap and send it to her. Her next letter would be full of the joy of being able to pass these out.

Ivy's early letters had instilled in me an enthusiasm for the possibilities of affecting other people's lives and trying to change their circumstances. It fitted the strong sense I had developed about the importance of fairness and equality. I was already, annoyingly, addressing basic issues of equality in the family: Why was it that Oliver and Aubrey were allowed to stay up late and I had to go to bed at the same time as Henry and Adrian? Why were Oliver and Aubrey allowed to go to the circus and I was kept home with the two younger ones? As the only girl, I was well able to hold my own with my brothers and I liked to compete with them on every level, but as the middle child, I felt that being grouped with the younger two played to my disadvantage. To the amusement of our parents, I became an argumentative champion of equality issues, not just on my own narrow behalf but also on behalf of my two younger brothers. I rarely won, but this did not stop me revisiting arguments again and again.

As we grew older we visited our grandparents' house, Amana, almost daily. It contrasted with our own house, which, while big and tall, was, with its flow of patients, somewhat confining. Amana seemed huge to us, a low, spacious bungalow in its own grounds with a yard and garden and fields. We spent most of our time playing outside or talking down in the yard to John Timlin, who looked after Grandpa's horses and told us stories about the older days with carriages and horses. For a period of a decade around 1850, Amana had been owned by a Mr Ham, the engineer responsible for the town's fine upper bridge over the river, known as Ham Bridge. A daughter of the house, Elizabeth Ham, kept a diary about the activities of Ballina, a garrison town at that time. In it, she described the house as the only building in Mayo that was neither a castle nor a cottage.

I particularly liked to sneak into the house and find my grandfather, HC, in the smoke room. He was a small, dapper man, who was often pottering around the garden or, being rather deaf, listening to the radio at full volume. I think Grandpa found it strange at first that I sought out his company, as he was the sort of professional man of his time who had little idea how to make conversation with a small child. He would produce a tin of biscuits or Black Magic chocolates out of the 'Laughing Cavalier' cupboard in the smoke room (so called because of the inlaid portrait on its door), and then talk to me like an adult. I was full of interest in his stories about cases he had argued. It is difficult to explain to a child what exactly a lawyer does, but his stories gave me a real taste of concepts of justice, fairness, and rights.

My grandmother was beautiful and talented and regarded as saintly in the town of Ballina. Almost unheard of for a woman, she had been master of the North Mayo Harriers and rode to the hunt side-saddle in a magnificent black outfit. She ran a number of charities in the town, and her door was always open to travelling women, who were referred to as tinkers. These women, who were living in dire poverty with absolutely no question of receiving state aid, would call to the door, and my grandmother would hand out blankets and food and money. HC would growl a bit, and my mother did not really approve, but Nellie chose to live her Catholic faith in practice.

My father's maternal grandmother, Louise, known to us as Great Granny, survived into her nineties. She was a handsome, tall woman who tied her hair up elegantly, and wore long earrings. As children we occasionally visited her house and were quite frightened by it. A terraced, four-storey house, quite close to the railway station in Ballina, it was dimly lit and full of formal furniture, so that we imagined all sorts of monsters lurking in the shadows. When I was eight she died and I was furious that we – that is, Henry, Adrian, and I – were considered too young to attend her funeral. So the three of us went out to the garden and climbed up on a wall (on which we were never allowed) from which we could see the cathedral directly across the river. We watched the hearse arrive and saw everybody standing, waiting outside the church. When the coffin was lifted out of the hearse, I instructed my brothers to jump up and down and wave and shout, 'Goodbye, Great Granny!' which we did in high, childish voices. Apparently this was vaguely discernible from the cathedral, so that later my father scolded me and sent me to bed as punishment for having put my younger brothers at risk in jumping up and down on top of the wall. I was filled with that kind of outrage a young child can feel at the perceived injustice of things.

Two years later another climbing incident heralded a new phase in my life. One afternoon, while we were visiting my grandfather at Amana, I was climbing a particular tree in front of the house that had very interesting branches. My grandfather came out with his walking stick and spotted me quite high up. He shouted at me to come down in a voice that made me move to a higher branch. He whacked the tree with his stick and said, 'You have to be sent away to school, you are not becoming a young lady!' He went inside and rang my father. When I returned home with my brothers, my father called me in to his surgery and told me that I was going to boarding school in September.

2

In Transition

To show your true ability is always, in a sense, to surpass the limits of your ability, to go a little beyond them: to dare, to seek, to invent; it is at such a moment that new talents are revealed, discovered, and realised.

Simone de Beauvoir, *The Second Sex*

WHEN I travelled with my mother to Mount Anville, the secondary school for girls in Dublin run by the Sacred Heart order, I was embarrassed to see that she was crying at the wheel and as she ushered me in to the Reverend Mother's office, whereas I was excited beyond belief and wishing that she would hurry up and take her leave. All summer I had been counting the days. I loved packing my new suitcase, and the tuck box full of goodies. By the time I left, my two older brothers had already returned to their Jesuit boarding school, Clongowes Wood. I felt important saying goodbye to the younger two. Just for a moment, as my father hugged me close, I felt a tear, but the new horizons that beckoned were too exciting for me to feel lonely.

Although I was still only ten, after the Reverend Mother had taken a look at me she decided to put me into the senior school, as I was tall and achieving good grades. This made me feel even more important.

Nonetheless, my early days in Mount Anville were not easy; getting used to the customs and rules took time. Early in my first year there, one of the older nuns stopped me in the corridor where I was walking with my hands in my pockets, and whistling. She said to me, 'Oh, don't do that, little one. Whistling makes Our Lady cry.' Even as a ten-year-old I knew this was a bit absurd.

Returning to school after my first Christmas at home was a more lonely experience. But I enjoyed my years at Mount Anville, doing well in exams and playing in the school hockey and tennis teams.

There were several other Marys in my class, so I was 'Mary B', and number 133. I was the youngest in the class, and now enjoyed the experience of having only girls as friends. We would gossip at night between cubicles, exchanging confidences.

The nuns in Mount Anville began to open our eyes to poverty in the developing world. We were encouraged by the Church to support the religious conversion and economic development of Africa's poorest infants, which ultimately amounted to buying black babies. I was enthusiastic about this, and spent all my pocket money buying these infants and assigning them dramatic saints' names like Aloysius and Alfonsus, right down the alphabet. It gave me a heady sense of power that I was able to 'save' so many of these black babies somewhere in Africa. Only later did I appreciate the underlying racism and appalling nature of this exercise that had absorbed my pocket money and my attention.

I was already a bookworm before I came to Mount Anville, but there I could satisfy my interests. I spent hours in the library once I became a senior student at fifteen and was permitted to go there after study and read until bedtime. I read the classics. I found books on Michael Davitt, Ghandi, Martin Luther King Jr., and Eleanor Roosevelt and became influenced by that sense of transforming the circumstances around you. I would sit on the windowsill in the library and daydream of being in difficult places and saving the world. I fantasised about doing something worthwhile with my life, based on the kind of work my aunt Ivy described in her letters.

I became friendly with one nun in particular, Sister Stephenson, whom we called Steve, who taught us mathematics and Latin. She could see I was hungry for serious conversation. I noticed that she had the *Irish Times* in her study. In my final year, aged sixteen – I had just turned seventeen when I sat the leaving exam – students were allowed to read only the *Irish Independent*, – a good, Catholic news-paper. When I questioned this, Steve quietly allowed me to read her copy of the *Irish Times*. I did not try to ensure that other classmates could have the same privilege; looking back, I see that once my own intellectual curiosity had been satisfied, I was content. The impulse to act on a collective behalf would come later.

During the summer holidays, when I was sixteen, I went to the Yeats International Summer School in County Sligo, as did my school friend Mary Courtney. We slept on rough horsehair mattresses in a dormitory of Sligo Grammar School and told one another ghost stories. We also attended classes taught by Professor Thomas Henn and other leading Yeats scholars. The Yeats Summer School instilled in me a love of poetry that has endured. It exposed me at this impressionable age to Yeats's poetry, and his plays influenced by the Japanese Noh drama, and to other great Irish writers, including Oscar Wilde with his savage wit and the desperate sadness of 'The Ballad of Reading Gaol'. I would have liked to be a poet or writer; I envy the ability to express ideas compellingly and economically. Throughout my career, I have drawn on poetry to convey points that in my own words would have been long and cumbersome. And privately, I turn to poetry if I am in need of inspiration. The book shelves nearest to my bed are where I keep a range of poetry books, so that they are easy to pick up at random.

In Mount Anville we were awakened at seven o'clock each morning with a bell, and by eight o'clock we had presented ourselves for daily Mass, wearing white veils and sitting in set places. I enjoyed the hour of Mass and prayers, a silent time of thought and, in effect, meditation. I was impressed by the nuns who had risen many hours earlier to begin their prayers and who also prayed late at night. The Sacred Heart nuns were not demonstrative but they were highly principled.

In our final year, almost all the girls in my class were discussing what they would do before they married. Marriage was their real objective. I, however, had no interest in it. I wanted to achieve something significant in people's lives, as Aunt Ivy had. I began seriously to consider becoming a nun.

Ivy was not the only formidable nun in the family. My father's paternal grandmother, Jane Bourke, was a Morrogh from County Cork, whose cousin Agnes Morrogh-Bernard, Mother Arsenius, came to nearby Foxford in about 1892 and founded the Providence Woollen Mills as a way of tackling dire poverty and unemployment in the area. This was social entrepreneurship of a high order, involving collaboration with an Ulster businessman, a loyal member of the Protestant Orange Order, who willingly provided the expertise required in setting up the looms.

A more powerful influence was my great-aunt Kitty, Mother Aquinas. She had been a distinguished presence in Britain as the Reverend Mother General of the Jesus and Mary Order. If half the stories we were told of how she bossed successive secretaries of state for education are true, she must have been a powerful ally or adversary. She had deep brown, twinkling eyes, had been quite a beauty in her youth, displayed a strong, feisty personality, and was terrific fun. In my teens I would go to visit her in the convent she had established, Gortnor Abbey, near Crossmolina, a few miles from Ballina. She had a special place in our hearts because she always had personal and affirming words for my brothers and me when we came to visit.

I was walking with Aquinas one day in the farm of Gortnor Abbey when we heard the hens in their henhouse. She shook her head sadly and said, 'You know, I often pity those poor cloistered hens, what it must be like.' I looked at her, and we both laughed. I said nothing but wondered if she was trying to convey a subtle message. Nonetheless, I took from her a sense that becoming a nun was a vocation for a strong-minded woman who wished to have an influence on the world around her.

Aquinas lived into her nineties, and we maintained a strong friendship well into my adult life. One thing I noticed in particular was that, like my mother, Aquinas would put her arms around us and, laughing, give us the warmest hugs. The Sacred Heart nuns did not show physical affection in that way. They were stricter; they never ate with us, and they did not hug or kiss us girls. They were affectionate in words, and shook hands warmly, but that was it. Noting a difference between the Jesus and Mary nuns who were happy to eat with you and show affection, and the Sacred Heart nuns, who were inspirational and had high ideals but who lived a separate, cloistered life, I wondered which was the higher form of vocation.

Preparing for my Leaving Certificate, I was aware that I had a good chance of getting high honours in a number of subjects. I had held on to the desire to travel the world and make a difference, and the more I considered it, the more I believed that the best way to achieve this, given my own personal faith, was to become a nun. As I saw it, my other options were to marry or to follow a creative path and become a writer. I chose to offer myself as a candidate to the Sacred Heart order because of the high ideals they lived by, even though I could see that they might benefit from the warmth shown by the nuns of the Jesus and Mary Order.

I went to see the Reverend Mother Provincial, who was visiting the convent. She was sitting behind a desk in the study when I came in. She asked me to sit, and listened as I told her that I had decided I wanted to become a nun. She said, 'Mary you're a bright girl, but I'm not sure that you're really cut out for religious life.' I was quite startled by this. I had assumed that once I informed the Mother Provincial that I wanted to be a nun she would be enthusiastic. Instead she looked at me quite shrewdly and said, 'I'll tell you what, go away for a year, and if at the end of it you still want to be a nun, come back and see me, and we'll talk about it.' I left her study puzzled and feeling rejected. I wondered what it was that she saw in me that did not make me good material to be a nun. I determined to prove myself to her.

My parents had been pleased and proud when I told them I wanted to be a nun, but, because I was just seventeen, they were relieved the decision was being postponed. Wanting the very best for me, they decided to send me to a finishing school in Paris for a year, and arranged for me to stay in a convent nearby and for friends of theirs living in Paris to keep an eye on me.

The year I spent in Paris in 1961, at the age of seventeen, changed my life profoundly.

I was met at Charles de Gaulle airport by Stuart Latta and his fiancée, Monica. Stuart's parents, friends and patients of my father, owned Massbrook House, an old fishing lodge on the shores of Lough Conn, about a dozen miles from Ballina (and, by great coincidence, the family home that my husband and I would acquire in 1994). A few weeks later I was a bridesmaid at their wedding. I did not tell my parents about this at the time. It was a Lutheran wedding, and in those days, the 1960s, Catholics did not enter Protestant churches or participate in Protestant services. I dithered, as I felt conflicted. According to what I had been taught, this was possibly a mortal sin. Nonetheless, I decided to attend and be bridesmaid but to refuse Communion and to pray as best I could during the ceremony. I did stop to wonder whether the Reverend Mother Provincial would think this further evidence of my not being cut out for the vocation of a nun, but only briefly. Before long there were far too many other novel and exciting things to occupy my attention. It was a sign of things to come.

On the evening I arrived in Paris, Stuart and Monica drove me to the convent of the Sacred Heart in rue Saint Dominique. We rang the bell, and the large imposing door opened automatically from inside. I entered, and a small, older sister wearing the familiar Sacred Heart nun's habit greeted me in French. That threw me. I had obtained honours in French a few months earlier, in the Leaving Certificate, but my French was learned from Moran's *French Grammar* and classical texts; it was not a living, spoken language. I could not understand what she was saying.

The sister showed me in and brought me to a room where I was to have supper by myself, as the other girls staying at the convent were out at a show. I sat down to a bowl of soup and some salad. There were two glasses on the table, one of which already had water in it. The sister appeared with a small carafe of red wine, which she proceeded to pour into my glass until I signalled refusal. I was an Irish Pioneer, part of a Catholic association promoting abstinence from alcohol; I wore a pin on my lapel that showed I had pledged, at my confirmation aged about eight, not to drink alcohol until I was at least twenty-one. I was stunned by the notion of a nun pouring wine – and a Sacred Heart nun at that! This was significant: nuns in Paris were allowed to drink wine. I knelt in the chapel later that night and reflected on my first impressions.

Mademoiselle Anita Pojninska's finishing school, known to all as 'Mademoiselle Anita's' was in the sixteenth arrondissement, two metro rides away. Quite often, though, I walked home, a seventeen-year-old in the streets of Paris dreaming and reflecting on the history of this incredible, romantic city.

At one level I was unimpressed with Mademoiselle Anita's emphasis on savoir faire or the appropriate manners and behaviour to exhibit in social settings; we were required, for example, to wear white gloves. But I was deeply appreciative of the quality of teaching there. Mademoiselle Anita employed professors from the Sorbonne who taught us as university students. We received a grounding in the history of philosophy in the way the French teach it – in a deep, conceptual way. We studied Hegel and Marx, Descartes and Sartre. We were taken to the Louvre and the Jeu de Paume and shown how to look at pictures; how to see works of art in the context of their time, and in the context of the medium the artist was using. I am not a visual person, but I was inspired and drawn out of myself. We were also taken to the opera, which I loved, both the performance and the extraordinary Second Empire building. Best of all, we saw and were utterly uplifted by Edith Piaf, 'the Little Sparrow', at the Olympia.

For the first term, I was in the English-speaking students' section. There I met the particular friend with whom I spent much of my free time in Paris. Cherry Richards was of a well-to-do Irish Canadian family that soon adopted me because, they said, I was staying in this rather gloomy convent and they had a fine modern apartment. Cherry was strikingly pretty, with blond hair and brilliant blue eyes under defined black eyebrows. She was a beautiful, happy, exuberant friend who, for some reason, took a strong liking to me, perhaps as her mirror opposite, a reserved, freckle-faced bookworm.

I studied hard and did well in my first term so that I was transferred to the French-speaking section of Mademoiselle Anita's, something that Cherry never aspired to. She would laugh at my interest in the French classics, the philosophers, my enthusiasm for Sartre and the novels of Françoise Sagan. Cherry became determined to distract me. She would drive to the convent on her scooter at about six in the evening to collect me. I would sit behind her and off we would go to the Latin Quarter, to drink coffee and talk. I found being with her a delightful change from the rather bookish and austere life I was leading. We would often go to the cinema in the evening, but on occasion we would skip classes and she would take me at ten in the morning to see films such as *Jules et Jim*. I would smoke the odd Gitane or Gauloise just to show how sophisticated and worldly I was. Early in my year in Paris, I remained a strict teeto-taller, but gradually the presence of the carafe of wine in the convent began to suggest there must be some pleasure in wine-drinking, and before the year was out I was enjoying a glass with the rest.

Cherry was a generous friend and wanted to buy me things. She lent me glamorous things to wear: skirts, blouses, and accessories. She taught me how to put on make-up. I felt more attractive when I was with her. She brought out in me a feminine quality that I had previously avoided. By then I had lost the plumpness of my child-hood, and had grown tall and slim. It was fun dressing up with Cherry and looking at myself in the mirror wearing fine clothes and lipstick.

This was the lighter, fun side of that year. But for much of the time I was miserable, lonely and plumbing personal depths I would rather have been spared. Arriving as a nun-in-waiting who was to spend a year in Paris before proving to the Reverend Mother Provincial that, after all, I was a good candidate for the order, I grad-ually became aware of frustrations that had been building up in me in a subliminal way for most of my life. I was asking questions of the world as I found it. Why did only boys became altar boys? Why did I have to wear a scarf in church when my brothers did not? Why did women have to wear hats and sit or walk behind the men? Why were the priests, the bishops, and the Pope himself all men? Why was it that women were encouraged to think only of marriage? Even *Bunreacht na hÉireann*, the Irish Constitution, decreed that 'the place of the woman is in the home'. Why was the emphasis so much on prohibition and fear of punishment? These were questions I had hardly dared ask in Ireland, but they came bubbling to the surface in Paris.

There were other matters that compelled my attention. I was introduced to notions of feminism, existentialism and secularism. I was becoming aware of a multicultural diversity that I had not been exposed to in Ireland and began to read left-wing journals and to watch radical films. I read about the horrors of the Algerian War and sympathised with the Algerian freedom fighters who were struggling against French colonial power. Everyone in Paris had to carry identity papers. I knew, being a white female, that I would not be challenged – however, Muslim women, girls my age, were being stopped and, even more so, boys who looked as if they might come from the Middle East. Many of the newspapers, of course, took the French side entirely and were in favour of repressing the rebellion, but not so some of the left-wing papers, particularly those circulating among students, whose articles quoted people like Che Guevara. I was troubled by the extent of the violence: the killing on both sides. Though my sympathies were with the Algerians, they, too, were responsible for terrible bloodletting.

I was more persuaded by the teachings of Ghandi and Martin Luther King Jr. and the Irishmen Daniel O'Connell and Michael Davitt, all of whom had espoused non-violence. This seemed to me to be a crucially important element in challenging the status quo. Whatever else, it should be done without violence to others. I went back to reading about Ghandi, particularly his period in South Africa, when he was a lawyer and led protests right up to the point where the protesters were beaten and stoned and he was imprisoned. His ability to persuade the protesting South Africans not to respond physically but to use their moral force was compelling. I read more of Martin Luther King Jr., and was deeply impressed. I saw a connection with writers like Albert Camus and his sense of the outsider: the observer who could analyse and clarify the justice of the cause, and bring about a change either by law or by moral force of public expression.

I also began to question the subtle violence of keeping one sex in its box. It seemed to me that my potential and my horizon were limited because I was female. I began to see that French women like Simone de Beauvoir, Françoise Sagan, and others were not in the same box at all. They had been liberated internally. I liked the idea of being free to think and do what seemed right from a personal, deeply-felt sense of integrity.

When attending Mass, I found the atmosphere in the chapel oppressive. On occasion I went to Notre Dame. It is a magnificent cathedral, but I found myself even less enamoured there. Tourists always circled the cathedral during ceremonies, but more than that, I was becoming increasingly sceptical: questioning the exclusively male priests and altar boys, and the very ethos of the Church.

I cannot identify any moment of a break with the past, but I had embarked on a different journey. I still embraced the idealism of Christian teaching: the importance of loving your neighbour as you love yourself, the importance of the concept that God is love, the wisdom implied in the Sermon on the Mount. But the change in me was fundamental. Now, instead of accepting the way of life that I had been brought up in, I was questioning everything around me in a constructive but nonetheless implacable way. Nothing was to be taken for granted, everything was to be examined and assessed. I did not feel this was being arrogant; rather, it felt essential as I sought an equilibrium within myself. From time to time, I still talked to a God; I felt the world made more sense if there was a God, but at the same time I did not appreciate how the Church represented this wisdom.

I became aware of the work of the theologian Hans Küng, who was developing a global ethic, drawing on the essential doctrines and values of the great religions of the world. I was excited at the idea that we could all renew our commitment through principles that could be shared and supported by all nations, cultures, and religions: a universal ethic as the basis for human rights, human responsibilities, and human connectedness.★ I felt this was the way to counter religious hatred – Catholic against Protestant in Northern Ireland, Hindu against Muslim, anti-Semitism – and xenophobia. I needed to know more about how societies worked in practice, so my mind was made up to follow my older brothers to university, where they were studying medicine, at Trinity but I resolved to study law.

By the time my father came to Paris to celebrate my completion of the year at Mademoiselle Anita's, I had become a very different

---

★ Years later, I would contribute my thoughts on this as president of Ireland to a publication, *Yes to a Global Ethic* edited by Professor Küng.

daughter. But if we were going to enjoy the week of holiday together, I knew this was not the time to broach the subject of my religious questioning. I was so happy to see him that I just wanted to spend time with him and not enter into any argument or discussion.

My father was oblivious to any change in me, and we had a wonderful week together. He took me to restaurants I would not have been able to afford. One evening while we were walking along the river the lights of the Eiffel Tower went out. My father turned to me and said, 'Mary, what has happened, did we do something?' Then we realised that it was exactly midnight, and we laughed together at the wonder of it all.

We took the *bateau mouche* on the Seine, we went to Mass in Notre Dame cathedral, which I attended dutifully, revealing no difference in my beliefs. I regaled him with stories of my time in Mademoiselle Anita's. We walked along the quays and looked at the books and pictures for sale. We visited exhibitions in the Louvre and the Jeu de Paume; we went to Versailles. He was inordinately proud that I had done well in Mademoiselle Anita's, had attained honours and had fulfilled his expectations.

I came back, however, to a Ballina that I looked at afresh. I still felt an affection for its small, intimate environment, but with a key difference. It was time to tell my parents that I no longer wanted to become a nun and did not wish to attend Mass. I completely underestimated their incredibly strong response. My father was furious, and my mother was extremely sad and inclined to weep. To them, declining Mass meant eternal damnation. My brothers were away – Oliver and Aubrey at Trinity, Henry and Adrian at Clongowes Wood boarding school – so I was alone in Ballina with my parents. I tried to reason with them; we had many days of argument. My father made it clear that as far as he was concerned, the only solution was for me to go to Mass daily and to speak to one of the local priests. I was too old to be smacked with the rubber knife, but I felt the full force of my parents' anger.

It is difficult to explain how isolated and helpless I felt in the face of my parents' utter opposition to this path I wanted to take, and their denial that I could possibly renounce Mass-going and adherence to the Catholic Church. I explained that I still upheld the principles of

Christianity, its standard and its moral compass, but this only made them angrier, especially my father. They couldn't countenance a departure from the Church's teaching. After a while I concluded that there was no way we were going to reconcile and be able to accommodate one another's views, so I subsumed my new beliefs, pretending, in effect, to my parents that I had somehow recovered my full faith. I was not proud of this approach, but in a sense I was trapped. My parents decided to delay for a year my going to Trinity. I think they thought that I would settle back into life as it had been and would thereby regain my old self. I faced a year at home, mostly alone with my parents while my brothers were away. The reality was that I would have to attend Mass regularly and show no independence of thinking.

I decided to work for an entrance scholarship to Trinity, partly to have something academic to do and partly because I liked the idea of having some economic independence. My interest in law had grown, prompted by those early talks with HC, my grandfather – about justice and about his taking on cases that protected the weaker party and brought about greater fairness – and compounded by my studies and reflections in Paris.

Sadly, I could no longer discuss things with HC. I had been shattered when he died towards the end of my time in Paris, in May 1962. I had not been allowed home for his funeral, as the flight was too expensive. This had been one of the loneliest times in my life. I had shut myself in my room in rue Saint Dominique for the best part of a week and did not eat. I went into a deep mourning, which I think was also in part a mourning for the religious life that had meant so much to me and that had been forever changed into a lonely, questioning path.

When I returned to Ballina, I often went to visit Amana. I wandered around the empty house trying to evoke that sense of my grandfather, of his habits, of the way he would go to the 'Laughing Cavalier' press in the smoke room and take from it treats for us children. When I opened the press myself, I found a three-quarters-empty box of Black Magic chocolates, and I wept.

Walking over to Amana with my father's Jack Russell terrier, Crumbs, became my escape. We would wander around, sit in the smoke room, go outside and down the yard, and I would think of the past, of the happiness of playing in the grounds, of walking around with my grandfather to admire the flowers, admire the vegetables and the gooseberries and raspberries growing at the end of the garden. Now the garden was going to seed, even though my mother tried to keep it up. She put my brothers and me to work in frenzied weeding fits. But mostly I was in Amana all on my own, feeling sorry for myself in a French-style depression that reflected the influence of Françoise Sagan and the existential novels I had been reading.

One afternoon, I was lying on the couch in the smoke room, crying quietly in a rather self-indulgent way, when Crumbs came over and licked my cheeks. Somehow that helped shake me out of the nostalgic stupor that was overwhelming me. I had started to find a sense of who I was. I decided to endure this year of being isolated from the excitements of Paris, helped by the prospects of a future life at university. I was glad that I was not going to become a nun, that I had acquired a taste for wine and had been exposed to new thinking and ideas. On the outside, I was obedient to my parents, going to Mass dutifully; at the same time, I built up an inner world where I learned to hide my true thoughts and feelings in order to create space for what I really wanted to think about and work out for myself.

3
Finding My Voice

In matters of truth and justice, there is no difference between large and small problems, for issues concerning the treatment of people are all the same.
Albert Einstein, undelivered speech, April 1955

THAT YEAR I spent at home was a tough, but in its own way formative, one. At the end of it I felt stronger somehow, more self-reliant. My need to find my own inner path, while outwardly following the weekly rituals of Mass and Communion, created some distance from my parents and resulted in my not needing their approval in the same way as before. Studying on my own, applying myself in that way, was a good preparation for university, and I looked forward already to the intellectual stimulus – and social life – that would come with being a college student.

In the spring of 1963 I sat the entrance exam and won a scholarship to Trinity College Dublin. My parents were proud of my achievement, though I endured great teasing about the general knowledge component of the paper. I had written eloquently on growing tomatoes and potatoes and on various other subjects about which I knew absolutely nothing, except how to bluff extravagantly. The scholarship provided basic financial support for the first two years of college. It meant a lot to me that I had some economic independence and could make a contribution towards the cost of becoming a law student.

I could not go to Protestant Trinity as a Catholic without the express permission of the then-Catholic archbishop of Dublin, Dr John Charles McQuaid, or else it would be a mortal sin on my part! My uncle Paget had been to Trinity, as had my aunt Ivy, before joining the Sacred Heart nuns and departing to India, and numerous cousins would be my college contemporaries. My father had visited the Archbishop twice already, to get permission for Oliver and Aubrey, and he was happy to do the same for me and for Henry, who was also coming up to college that year. He assured me that the visits were painless, and that the Archbishop offered him a glass of sherry on each occasion.

I realised the power game that was in play – the Archbishop was keen to deter prominent Dublin Catholics from sending their sons and daughters to Protestant Trinity rather than to University College Dublin (UCD), the university linked to Cardinal Newman. (The pressure applied to country Catholics such as my parents was evidently less.) I kept my scornful feelings to myself, and smiled quietly when, a year later, the Archbishop decreed that it would no longer be necessary for any Catholic to obtain permission to attend Trinity. I did tease my father gently about not getting his glass of sherry when Adrian followed us to college.

Oliver and Aubrey had used up the residency time students were allocated to live in rooms at Trinity, and my parents needed to find accommodation for all four of us in Dublin. My mother shrewdly purchased a house, No 21 Westland Row, right at the back of the Trinity campus, in an executor's sale. Actors Micheál Mac Liammóir and Hilton Edwards had unveiled a plaque by the front door, in 1957, to mark it as Oscar Wilde's birthplace. So the four of us – Adrian made it five the following year – were students together and occupied Westland Row with Nanny Coyne. Far from being inhibiting, Nanny's presence with us in that house was incredible. She cooked for us, cleaned for us, looked after us, and put up with a great deal of partying and sheer student exuberance. She also took a huge interest in what we were all doing, and read the books we were reading. She was a friend and ally, and when she became aware that I had stopped going to Mass regularly, notwithstanding her own unshakeable faith, she was discreet and took my side. She appreciated that this decision was deeply felt on my part.

By the time I arrived at Trinity, Ollie and Aubrey had amassed a prodigious number of friends. They were highly social and, following in the footsteps of our mother, enjoyed student life to the full while comfortably getting through their medical studies. Aubrey in particular had a lazy intelligence, an easy charm. He was my beloved older brother, who was just delightful and life-enhancing to be with. This made him popular with a wide circle of friends, as did his innate exuberance and generosity. Needless to say, he was much pursued by the opposite sex. He was a gifted sportsman, captaining the Trinity rugby team, and capped for Leinster and later for Connaught when he moved home to Mayo as a doctor. He also captained the Ballina rugby team, as well as playing Gaelic football.

Ollie was also a talented rugby player and played for Trinity and Wanderers; he later coached the Irish under-nineteen team with distinction, making lifelong friends with some of his charges. He and Aubrey had a great interest in horse racing, and when Henry and Adrian arrived at Trinity, they, too, embraced the social aspect of college with vigour. The impromptu parties my brothers threw in Westland Row tended to be memorable and could get quite messy. On one occasion, Ollie and Aubrey had managed to lay their hands on a large quantity of drink: vodka, gin, rum. They put the stopper in the bath and emptied bottle after bottle in, mixing an enormous cocktail, which they scooped out into buckets, and proceeded to distribute in glasses to the guests. It seemed as if the world and his wife turned up to this party; there was a queue down the street in Westland Row. In the early hours of the morning, one eminent professor was found fully clothed in the bath, slopping around happily.

I tended not to be involved in these parties, but I enjoyed the fun and gossip. I also loved to tease my brothers that the shop of the local bookmaker, Harry Barry, which had been run down when we moved in to Westland Row, was now completely renovated, clearly, I said, with the profits from their pocket money. Sometimes I joined in with my brothers' exuberance, on one occasion gatecrashing the Trinity Ball, which necessitated climbing a ladder over the back wall in a full-length dress. Henry held the ladder as I went up, and a grinning Aubrey sat on the wall, coaxing me and helping me land safely on the other side. Meeting up with friends and dancing to the bands, we roamed gleefully until dawn.

Our mother, who was a regular visitor, arriving at Westland Row with bags of vegetables from the garden in Ballina, was somewhat torn: on one hand she shared my father's strict disciplinarian approach to parenting, but on the other she was brimming with love for the social aspects of life and her own fond memories of student days. On one occasion, in Freshers' Week, when many of the university societies enrolled new students, Henry and I went to one of their parties and drank just a little too much of the cheap, sweet South African sherry that was on offer. We stumbled home somewhat the worse for wear, and the next morning our mother, on one of her overnight stays, greeted us with a miserable face and started saying, 'I cried myself to sleep last night,' but then almost immediately broke into giggles, as it was the opening line of a popular song.

My friend Mary Courtney encouraged me, as did others, to explore the social life they were enjoying at UCD, which was rarely done at the time by Trinity students. I went to meetings of the UCD Laurentian Society, a Catholic society, where we debated the role of religion. Some, like me, were all for change; others were more conservative. Surprisingly, the UCD chaplain, Father Jack Kelly, a Jesuit, was one of the more progressive voices. I became friendly with and confided in him, and was immensely relieved when he encouraged me to follow my own path in seeking truth, and not to feel guilty about no longer attending Mass regularly. He gave me space to breathe, which liberated me and lifted a burden. There were few priests at the time with that ability to encourage a questioning spirituality. Generally the orthodoxy was absolutely unshakeable; there was little room to follow one's own path. For that reason I was slow to confide, but I did find another guide, Father Enda McDonagh, a professor and moral theologian in Maynooth College, who became a lifelong friend.

Because of the year I had spent in a lonely introvert's circumstances in Ballina on my return from Paris, I felt older, more mature, and more appreciative when I came to university. Dublin in the 1960s was alert to what was going on in the rest of the world – its fashion and music scenes, its mood of change – and was following suit. I found that listening to artists such as Bob Dylan and Joan Baez suited my spirit of questioning the status quo. RTÉ, Ireland's national broadcaster, had begun transmitting in the early sixties, bringing much more awareness of international news and cultural happenings, and the *Late Late Show* was already exploring topical issues and beginning to drive social change. Arriving in this Dublin, once again I was in an intellectual environment where I could be myself. I studied law with the enthusiasm of having waited an extra year to get there. There were twenty-eight students in our class, of whom four were women. At that time, students in 'legal science' – the rather archaic name for the four-year undergraduate law degree at Trinity College, which was considered to be on a par with the teaching of law at Oxford and Cambridge – were required to take a subsidiary general studies subject. I chose to take French, and decided that instead of taking pass French, I would take the honours course. This gave me the opportunity to continue to study the French classics and, significantly, to be taught by Owen Sheehy-Skeffington, then a university senator, a first cousin of the diplomat and scholar Conor Cruise O'Brien, and well-known in liberal circles for the causes he espoused. I admired Owen's outspokenness; he was a frequent writer of letters and articles with caustic comments on the Catholic political culture of the time.

Most law teachers tended to lecture from their notes. The skill, for the students, was to take the notes down quickly, learn them, and more or less repeat them in examinations. I was particularly good at this, and worked hard to achieve good results. But it was a bit dreary as a teaching method, with little interaction or student input. The Socratic method of law teaching was unknown to us at the time, but I felt – and this was based on the exposure I had had to the teaching by Sorbonne professors at Mademoiselle Anita's – that there could and should be more to the teaching of law than mere dictation of notes. I had an inkling, even then, that I might teach law.

My brother Henry, having opted for a pass degree in general studies, took a very different approach to his first year at Trinity; he enjoyed student life so much that he managed to fail the almost unfailable subject. I couldn't resist teasing him, telling him that this was truly quite an accomplishment! However, it was also serious, as it meant he might have to drop out of university altogether. My advocacy skills came to bear in persuading our shared tutor, the economist Alan Tait, that Henry was actually somebody who should have studied law. Law being an honours subject, and therefore more difficult than general studies, it was a notable achievement to succeed in persuading Tait to support Henry's switch. My judgment was borne out when Henry went on to become a formidable senior counsel practising on the Western Circuit.

I enjoyed Kader Asmal's lectures on administrative and public international law, although some students would mutter that he never used one hundred words where five hundred would do. Kader was very accessible to students, and spoke with passion of why he had had to leave his native South Africa.*

The first-year results showed that three law students achieved first-class honours: Don Beck, Nick Robinson, and I. I was aware of Nick and mildly friendly with him before that, but our going out to celebrate our results together was the beginning of quite a friendship, although at that stage there was no thought on either side of developing it further. Nick was handsome and intelligent, with an artistic, creative sensibility. I would see him accompanying some of the prettiest women in college. He designed and drew posters for the various debating societies and college events, and this earned him a steady income. I particularly enjoyed his sense of humour, and that he was comfortable and supportive in female company. He lived at the time with his aunt Audrey, his father's sister, in Monkstown. His mother had died when he was very young, and Nick and his three brothers had been raised by their father, Howard Robinson, a successful chartered accountant and entrepreneur, and a prominent member of the Anglican Church of Ireland, sitting on its representative body and advising on its finances.

From time to time Nick invited me for dinner at Bernardo's restaurant, just outside one of the side gates of college and around

---

* Later I joined him in the anti-Apartheid movement he and his wife, Louise, established, and served with him when he founded and chaired the Irish Council for Civil Liberties; we became close friends. Later still I would celebrate with him the new South Africa he was helping to shape – so inconceivable at that time.

the corner from Westland Row. This was a real treat, and led to our becoming increasingly comfortable in each other's company. I also became friendly with the UCD law student Fergus Armstrong, and Nick, Fergus, and I co-edited a short-lived undergraduate magazine called *Justice*, for which Nick also drew cartoons. It was a rare example at the time of inter-university cooperation.

I was particularly interested in constitutional and legal theory and by this time had decided that I wanted to practise my law as a barrister – a litigator – which would involve arguing in court. Being a woman, I would have to be that bit better, work that bit harder so as to argue with utter confidence. There is no room for timidity in litigation. So I forced myself to take part in debating, going to UCD frequently, especially to the L&H, the Literary and Historical Society, UCD's primary debating society. (Trinity's historical society, 'the Hist', still barred women.) Despite having prepared meticulously for these debates, I was prone, initially, to stage fright, but gradually I began to be able to hold onto the logical thread of the argument and relish the exercise of advocacy. I found within myself a competitive instinct and a determination to articulate my point. Harry Whelehan, a friend and fellow student at King's Inns, where we studied for our professional Barrister-at-Law degree in order to practise law, was a clever, relaxed debater, and we teamed up and won a debating competition held in Cork. This did wonders for my confidence, because it was rare for a female student to participate in, let alone win, such a competition. I felt the rush of adrenaline, the sense of being appreciated by a live audience – and I liked it.

I decided in my second year in college to attempt the scholarship exam. A scholarship in legal science was thought to be particularly difficult, so much so that people rarely sat the exams, and it had been ten years since anyone had got 'schol' in law. Nonetheless, I relished the challenge, studied hard, and won a scholarship. This meant financial assistance for the rest of my time as a student in Trinity, and the status of being a 'Trinity scholar'. In those days under the archaic statutes of the college, male scholars were considered part of the foundation of the college, while female students had to settle for being 'non-foundation' scholars, with fewer privileges. I was annoyed at this, but it was so much a part of the culture of the time that I didn't kick up a fuss – there would be other battles.

During this time, I was constantly developing my sense of the law's potential as an instrument for the social change so needed in Ireland. The Irish Constitution and laws reflected the ethos of the Catholic Church, whose predominant position in society was guaranteed in Article 44. The effect was to marginalise those of the Protestant tradition, the small Jewish community and those who were not religious. This 'special relationship' (as the Constitution then put it) was a huge barrier to greater understanding between north and south in Ireland. Trinity had a strong Protestant ethos and attracted students from England, many of whom saw it as an alternative to Oxford or Cambridge, as well as students from the Protestant tradition in Northern Ireland. Classmates and friends from Northern Ireland expressed their feeling of being constrained because of the exclusive nature of the Catholic ethos in the Irish laws. I felt that Ireland needed to open up and be willing to value diversity.

At the same time, both in Trinity and UCD we were actively discussing and debating the blatant discrimination against Catholics in Northern Ireland, 'the North' as many preferred to call it. This extended right across the board – to housing, education, jobs – and was accompanied by harassment by the police, the Royal Ulster Constabulary (RUC). It was easy to identify with the pent-up frustration that would lead to the civil rights movement in 1968.

My particular friend at Trinity was Eavan Boland. I met Eavan – who at the age of nineteen had already published an acclaimed book of poetry – in the spring of our third year. She was the daughter of the then-chancellor of the university, Frederick Boland, an eminent Irish diplomat, well known for his chairing of the UN General Assembly. (Freddie famously broke his gavel trying to silence Soviet president Nikita Khrushchev.) Her mother, Judy, was an accomplished artist.

Once, during a break from lectures, while we were both having coffee in the Buttery, the dingy student café bar located in the basement below the dining hall, we fell into a conversation that, in one way or another, has been going on ever since. I invited her to come to lunch in Westland Row. As we walked, Eavan told me about some academic difficulties she was having – though she would go on to achieve a first that year, and the next. I wanted to impress upon her that she had the poet's imagination, that this mattered more than academic rank or status. Later, at lunch we quickly found common ground comparing our lives; she had attended Holy Child, Killiney while I was at Mount Anville. Her brother, Fergal, had been at Clongowes with my brother Oliver. She understood some of the pressures I had begun to resist, though, as a diplomat's daughter, moving from place to place, she had experienced different pressures. Both of us were in our own ways questioning society's approach towards women, seeking a set of values by which to live and work, formulating ways to express our uneasiness with society and its attitudes.

From then on we would meet regularly, walk up Grafton Street, and sit in a café, having these long conversations about poetry and law, the central role of the imagination. Eavan introduced me to other poets: Seamus Heaney, who was a bit older and already teaching; Paul Durcan, whom I knew from Ballina connections; and one of her lecturers, Brendan Kennelly. All three of them also became friends.

Eavan talked about how she felt isolated from the Dublin social set her older sister and brother were involved with. Her readiness to speak frankly allowed me to open up about my own family dynamics and my unwillingness to accept, without question, Church doctrine. These frank, intimate discussions bonded us. We became closer and closer, and even when we were parted for months, we simply resumed our conversation upon reconnection. Today we can be parted for years and do the same. Eavan, now a renowned poet and professor of poetry at Stanford University, helped me understand the value of true friendship and the closeness you can have to somebody when you are bonded in that kind of friendship.

One evening Eavan, who had joined the Women's Liberation Movement, rang me and asked if I could give her seven examples of laws that discriminated against women for a paper she was working on for the movement. 'Only seven?' I said. I could think of many more. But she said that seven was a good number and besides, she had limited space. We both laughed, yet we were serious.

By then, I had a profound sense of the extent of discrimination against women in Irish society, and was deeply, unequivocally supportive of women's rights and the struggle for equality, but at that stage I felt restrained in using the language of feminism or in being openly identified with the feminist movement. Those who identified themselves as feminists, I felt, tended to be pigeonholed. I didn't want to limit myself to one issue and I didn't want, through being stereotyped, to be branded and isolated as a feminist. I could see, with almost dogmatic clarity, how changes could be brought about legislatively, from within the law. Eavan and many other women didn't agree. They believed that change could be brought about only through resistance. Therefore, I didn't get involved directly in front-line activism; I didn't, for example, participate in the much-publicised train trip to Belfast where women stocked up with contraceptives to bring back down South in open defiance of customs officers waiting in Dublin.

My approach was to identify the areas of law that needed to be changed and then figure out how to go about changing them. This was a tactical decision, and it was one I deployed throughout my career as a barrister and a senator: to be more effective and to gain more support by not being typecast and thereby limited in my capacity to effect reforms. Tackling discrimination against women was part of a broader need for law reform.

Trinity wasn't all work. I enjoyed the café life and going to the theatre, especially the Players Theatre in Trinity itself, which was rich in talent. I loved the social element to debating, meeting over a glass of wine with the others involved. I played tennis and hockey for Trinity and was generally good at sport, though I didn't give it the same level of commitment as my brothers did. A *Trinity News* photographer captured a dramatic moment in hockey when I scored a goal, my stick blatantly and illegally straight up in the air. I also enjoyed rowing. My three female colleagues in our legal science class and I trained with a ferocious male cox, and entered the ladies race in our second year. We called ourselves 'the Four Just Women'.

It did not help, as we limped home in last place, that some of our male classmates were running along the bank of the River Liffey waving a poster that proclaimed, 'Just Four Women'!

A few of my male contemporaries approached me at one time or another and asked me out, but I resisted, not ready to commit to anyone at that stage and not looking for such a relationship.

In my final year, the new Regius professor of law was a Canadian, Desmond Morton, who had a more challenging style of teaching. He welcomed interventions from students, which I responded to eagerly, taking the opportunity to raise the issue of travellers' rights. Now recognised as a distinct ethnic minority, the travelling community (also disparagingly known then as tinkers, because of their craft in making and repairing tin goods) is of a distinct and ancient tradition akin to gypsies or the Roma peoples. They tend towards a nomadic way of life and have a strong, clannish sense of family. At the time, many lived in caravans and travelled the country, stopping at sites according to tradition and season. As they did not receive any kind of state aid, many of them lived in circumstances of dire poverty and suffered discrimination and prejudice in Irish society. The policy of successive governments had been to attempt to assimilate them into the settled population.

I suggested that we address this issue. So Professor Morton decided to hold a seminar on travellers, but not, to the surprise of the class, in Trinity. We went instead to Cherry Orchard in West Dublin and visited a traveller halting site with its bleak conditions and single water tap for the thirty or so caravans. We then retired to a café nearby and sat around in a group, holding the seminar there. Morton asked us what the best way was to tackle the traveller issue. Not able to answer, we diverted the talk to the subject of begging. He pressed one idea strongly: nobody should give money to those travellers begging in the streets, because that was not a solution. Doing that, he said, allowed the state to get off the hook. Instead, the state should be required to provide proper water, proper sanitation, and aid. Professor Morton challenged us that this was both a state and a city responsibility. Our giving of a few shillings, which we were all doing, was actually not a good idea. What was required was a kind of 'tough love' approach. It was easy to give money to individuals, but what was really needed required more effort, and that was to come up with policies to address travellers' needs, and to shift traditions and attitudes. These people needed proper protection as citizens of Ireland. I remembered that seminar vividly, and later in my law career I would represent travellers suing the state in various cases for determination and vindication of their constitutional rights.

In my final year I ran for auditor of the Trinity Law Society, a small but prestigious (and male-dominated) undergraduate society in which I had been active as a debater. This was my first time seeking election. Once I was a candidate, my brothers and their friends joined in canvassing and securing votes, and I won the election, becoming, I think, the first woman auditor.

I took my role to heart, ensuring that I secured speakers of high quality, particularly from the United Kingdom, including the Master of the Rolls, Lord Denning, who gave a brilliant, quirky speech at the annual Law Society dinner. It was a black-tie affair with all the pomp and circumstance, but ended up with much singing and banter. I remember that Nick sang 'The Blackbird'. My brother Henry's girlfriend, Barbara MacKenzie, who would later become his wife, got up on the long table we were all sitting around and danced elegantly with a rose between her teeth.

To my surprise, the easy friendship with Nick that had developed over the years changed on his side to something much deeper. I remember the context but not the exact date. I had made it to the finals in a debating competition to be held at UCD. Nick offered to come along with me. When I stood up that night, all the earlier practice helped; I was passionate and eloquent, got a standing ovation, and won the individual award. And then Nick kissed me, and I could sense that he wanted a different relationship. I still wasn't ready to let anyone get that close, not even Nick. I was more focused on my inner world and on my studies. Nick accepted this, but told me he was serious and prepared to bide his time.

The preparation for my inaugural address as auditor of the Law Society was a challenge. I was determined to take on a subject that would capture the potential of law to be an instrument for social change. I recognised increasingly the need for Ireland to open up, the need to change the Constitution and laws to give space for private morality. I had been following the debate in the UK between Professor HLA Hart and the more conservative Lord Devlin on the relationship between law and morality. I chose for my speech the topic 'law and morality in Ireland'. Almost nothing had been written in Ireland on the subject, so I consulted Professor John Maurice Kelly of the law school in UCD. He was an outstanding authority on Roman law and Irish constitutional law, and I had admired some lectures of his that I had sat in on. When I outlined my proposed topic on law, morality, and the inappropriateness of some of the laws in Ireland, he looked at me disapprovingly: it was not, to his mind, a good lawyer's subject; there was very little law in it. If I really wanted to do a good inaugural paper, I should go back to the drawing board.

If I had in some unconscious way been seeking permission from Professor Kelly to pursue this line I realised then that I didn't have it, but despite being deflated, I felt strongly enough about my topic to persist. I discussed my ideas with Eavan and realised from her reaction that, although she was urging me on, I was treading on uneasy ground. I had identified moral issues such as the constitutional prohibition on divorce, the ban on the use of contraceptives, the criminalisation of homosexuality and suicide, and I wanted to make the case that these matters should not be criminalised by the state or reflect the morality of one dominant religion to the exclusion of others.

I went to see Justice Kingsmill Moore, a retired judge of the Supreme Court. Theodore Kingsmill Moore had kept up a strong connection with Trinity and made himself approachable to students. He was intrigued by my idea, I think, and supportive, letting me know that this was exactly the subject I should broach. When I told him I had been reading the debates between Professor Hart of Oxford and Lord Devlin, he suggested that I invite both over: Devlin sent his regrets but much to my surprise and delight, Hart accepted, as did Vincent Grogan SC, a skilled parliamentary draftsman and, at the time, Supreme Knight of the Knights of St Columbanus, a society of Catholic laymen based in Dublin. Kingsmill Moore also agreed to participate. It was quite a coup, that these distinguished authorities would take part in my student inaugural address, to represent both sides of the argument.

We hung posters (designed by Nick), all over college, and the topic seemed to capture the imagination of the student body, so that it was decided I would deliver the inaugural address in the Examination Hall. As far as I know, this had not happened before. The speech would normally be given in one of the student society rooms in college.

Both my parents were due to come to the address. I wasn't too worried about my mother's reaction; for her, the pleasure would be in seeing her daughter up on the stage, in the limelight. But my father would be more concerned with the content of my speech, so, looking for his blessing and approval, I discussed it briefly with him and gave him in advance a copy. Unbeknownst to me, he showed it to his friend the local bishop. Bishop McDonnell reassured him, taking the attitude that this was undergraduate thinking; it was what students did and should do. This gentle absolution gave my father peace of mind, though he said nothing to me about his meeting with the bishop until much later.

On the evening of the event, the hall was absolutely full. There was a terrific buzz of excitement. For a moment I felt a surge of anxiety and nervousness, remembering JM Kelly's admonition that mine was not an appropriate topic. What I was about to say was radical; I would be attacking some sacred cows, and I had no idea how my ideas would be received. As soon as I started speaking, the adrenaline helped me find my voice. I spoke about the 'special position' of the Roman Catholic Church in the Irish Constitution, and how the laws were enforcing Catholic morality, mistakenly equating 'sin' with 'crime'. I espoused removing from the Constitution the prohibition on divorce, lifting the ban on use of contraceptives, and decriminalising homosexuality and suicide, on the basis that these were personal moral issues that should not be subject to the law of the state but should be up to the individual to choose, based on his or her own moral or religious code. When I finished speaking there was a prolonged silence, and then came great applause. This was a seminal moment for me, a moment of validation. This was why I had studied so hard; this was what law was about. There was no overt dissension at the time; that would come later, as I sought to implement these ideas.

During my final year, through the good offices of Desmond Morton, I was invited to take the minutes of a conference in The Hague preparing a convention on private international law. While taking the minutes in English, I became friends with Cyril David (son of a well-known French professor), who took the minutes in French. Cyril and I would finish our work in the Peace Palace in The Hague at about eight in the evening, rush to his *Deux Chevaux* and drive quickly to Amsterdam to find an Indonesian restaurant that was cheap enough for us to enjoy a hearty meal and get back to The Hague by midnight. This was thrilling, cosmopolitan living. I was being paid seven pounds a day, which was a fortune. I relished the intellectual work, and got to know the chief US negotiator at the conference, Professor Arthur von Mehren. When he asked me what my plans were on graduation, I told him that I was seeking to do postgraduate work in either Oxford or Cambridge. He suggested that I consider the Harvard Law School, invited me to apply for a fellowship, and sent me the details when he returned to Harvard. This was an exciting prospect. I filled out the forms, sent them off, and was delighted to receive a letter offering me a fellowship.

In all, I took three degrees in 1967. During my final two years at university I had also been studying at the King's Inns for my BL, my professional Barrister-at-Law degree. At the same time, I took Trinity's LLB exams in June, and then in early September, I took its legal science finals for my BA. The class swot, I achieved first-class honours in each.

Nick continued to pursue me gently, but as I was going away to Harvard, I was not thinking about developing our relationship. On the morning I left Dublin to fly to Boston, he came to Westland Row bringing flowers for me. I had already left. He gave the flowers to a sympathetic Nanny, who, knowing my parents and how they had me on a pedestal, was not particularly encouraging of his prospects with me.

Nick stayed on at Trinity as an LLB student, and succeeded me as auditor of the Law Society. He worked hard at that, gaining the Vincent Delaney medal, and harder still on his burgeoning career as a political cartoonist, already submitting work to journals such as *Dublin Opinion* and *Hibernia*. I was delighted to learn of Nick's achievements, and we continued to exchange letters from time to time. I could tell that he was worried I might meet a handsome American law student who would sweep me off my feet. I had, indeed, moved on mentally to Harvard, with all its exciting prospects, but despite some overtures by Harvard contemporaries, I was not going to lose my heart to anyone during that year.

4

Broadening My Horizon

*An individual has not started living until he can rise above the narrow confines
of his individualistic concerns to the broader concerns of all humanity.*
                    Martin Luther King Jr., *The Measure of a Man*

THE TERM at Harvard started in the last week of August 1967. As the
Trinity finals for legal science were in early September, I literally put
down my pen from my last exam and caught a plane to Boston. I
arrived fairly exhausted, and had missed the first week or so of term.
My friend Fergus Armstrong, who had graduated from UCD, was
already there when I arrived, and he helped me get organised and
showed me around. I found the Harvard Book Store and was shocked
at how expensive the law books were. I would have to manage on
the fellowship and some additional support from my parents. I
quickly discovered that, whereas I usually went to breakfast at about
eight in the morning, early by Irish standards, most of the students
at Harvard breakfasted before seven. But from now on I would
do the same. America was going to be different.

Boston reminded me of Dublin, but with notable differences. I
loved walking along the Charles River, more open and accessible
than the Liffey in Dublin, and full of rowing boats and other pleasure craft until it froze in winter. I was astonished – and flattered – to
find that if I went to a pub and wanted to order alcohol, I had to
produce my passport or student card to prove I was over twenty-
one. The servings of food in restaurants seemed huge, and you were
offered a confusing variety of sauces with everything. It was disconcerting to see how little coverage there was of Europe and European
affairs, never mind Ireland, in the US newspapers. Once a week I
would buy the *Irish Times* in the Harvard Square newsagents, just to
have some news from home.

To my surprise, my fellow students all dressed casually, mainly in T-shirts, sweaters, and jeans. Thanks to my mother, I was one of the most fashionably (though formally) dressed, in skirts and blouses, dresses, and high heels, though I began to dress down quite quickly, and with some relief. As law in the US is a graduate subject, the students were in their mid- to late twenties. Quite a number of the men had wives who worked locally – as the joke went, doing their own 'PhT', or putting hubbie through. There were never more than ten female students in any class I attended, where the class size varied from fifty to about one hundred.

As I got to know the students from different parts of the United States, I realised how deeply many of them felt about the immorality of the war in Vietnam. I was relieved, impressed by their articulation of ideas, and shared my own views and criticisms. I was even more impressed when I learned that a number of those graduating that year were not looking for jobs in top law firms in New York or Washington but were applying for positions in poverty programmes, or wanted to be part of the civil rights movement to fight racism. It was an exciting, idealistic time, and I felt inspired by the courage and commitment of my fellow students.

The international students doing the LLM, the one-year masters degree in law, were invited early on in the academic year to dine at the residence of the dean of the Harvard Law School. Professor Erwin Griswold, who subsequently became solicitor general of the United States, was known as 'the Griz' and, living up to his nickname, was a grizzly, gruff guy. We were about twenty-five students, and after dinner he invited each of us, in alphabetical order, to introduce ourselves and say why we had come to Harvard and what we hoped to do. There were some 'A's, including Fergus Armstrong, who did a po-faced, serious introduction of themselves. By the time it got to B, for Bourke, not that far down, all the students before me had offered polite, earnest introductions, so I decided to lighten it up a bit. 'I'm Mary Bourke from Ireland. I'm here to have as enjoyable a time as possible and while I'm at it to see if I can get a good degree from Harvard.' This did not go down at all well. My fellow students looked aghast. Nobody laughed, and worse, the Griz fixed a beady eye on me, as if to say, 'I have your number'. Years later, I heard with amusement from the then-US ambassador to Ireland, John Moore, that his friend Griswold had regaled my comment with some enjoyment and had followed my ensuing career with interest.

From the start, the pressure of work in the law school was intensive. This was a highly competitive environment. No matter how late you worked in the library, there were always people working later; no matter how early you got up in the morning, there were lights on in people's dormitories, and you knew they were up and working.

On the other hand, it was very positive being Irish at Harvard and in Boston. While in the past Irish people simply could not get into Harvard – I thought of how Jack Kennedy had broken through the white Anglo–Saxon Protestant culture – now being Irish was looked on favourably. At the same time, there were lonely periods, when the work was hard and I felt far from home. Fergus Armstrong and I had agreed to set up a hot-line between us, for those moments when the pressure and loneliness became too much, and occasionally we would get together for a drink to commiserate.

I was in regular correspondence with both my parents, and hungry for news. Nick sent witty letters commenting on the gossip from college and Dublin social life, sometimes illustrated with his cartoons, which made me smile. About once a month, on a Sunday evening, my parents telephoned me. My mother would talk about family, what my brothers were up to, how Aubrey had scored a try for the rugby team. My father would ask a bit about how my studies were going and would then launch into a furlong-by-furlong account of the latest big horse race he had watched on television, how one horse had fallen at the last or another had come up on the inside to win the race by a head. I had little interest in horse racing, and what he said would go in one ear and out the other, but I loved hearing his voice and enthusiasm.

I found the teaching in Harvard inspirational. Most of the teachers used the Socratic method: you looked at the materials in advance of class and then you got tough questions in front of fellow students. We were required to read cases and then, in class, the professors would keep changing the parameters of the question, or altering the basic facts, so that you really had to think on your feet.

As well as doing my required number of LLM subjects, I also audited five or six other classes, not for credit but purely out of interest in the subject and how it was taught. I knew at this stage that I was definitely going to teach as well as practise law. I audited Labour Law with Professor Archibald Cox (who would later investigate President Nixon); Constitutional Law with Professor Paul Freund, at the end of whose course the students gave a standing ovation; I went across to audit an undergraduate course of the famous economist John Kenneth Galbraith, who discussed the economics of India, having been US ambassador there recently. He lectured in the old Trinity style, delivering his script in a wonderful voice while the students busily took notes. I was flattered by the interest he took in my studies.

Another course I audited was Law and Psychiatry, taught jointly by Professors Alan A. Stone and Alan Dershowitz, the celebrated trial lawyer. Professor Abe Chayes, who taught a course I was sitting in international law, had been in the Kennedy Administration during the Cuban Bay of Pigs affair and, as a case study, was giving us a blow-by-blow account of that recent controversy. I also audited a course on urban legal studies taught by Professor Adam Yarmolinsky, examining the fact that white families were moving out to the suburbs and leaving black and Hispanic ones in the cities, as a result of which the cities were becoming run down. I had never been taught law like this before. It felt cutting edge, and it made me appreciate that if law were taught in this challenging style, in a way that urged problem-solving, it would make a great difference in provoking a more thorough assessment of the issues involved.

While at Harvard I lived at Wyeth Hall, a comfortable women's dorm with great facilities. I shared rooms consisting of two bedrooms, a kitchenette, and bathroom with Hélène Le Bel, a Canadian student. Her father had been a highly regarded, idealistic labour lawyer who died young. Hélène and I became very good friends. I would have loved to go home to Ireland for the Christmas holidays, but the expense of travelling was just too much, so instead, at Hélène's invitation, I went to Montreal.

Hélène lived with her widowed mother and brother, Bernard, in quite poor circumstances; like me, she had won a fellowship to pay for her tuition at Harvard. French-speaking Canada was abuzz: a few months earlier, President de Gaulle of France had controversially declared from the balcony of Montreal City Hall, '*Vive le Québec libre!*' There was secessionist fervour, and I got caught up in it. We went to a concert hall where Gilles Vigneault sang, '*Mon pays ce n'est pas un pays c'est l'hiver*' (My country is not a country, it is winter), a terrifically emotional song. And then I spent Christmas itself with Cherry Richards (my friend from Mademoiselle Anita's in Paris), now married to lawyer Michael Haynes and living in Montreal. There was a polite greeting but little empathy between the French-speaking, intellectual, rather cerebral Le Bel family and the well-to-do, sociable, English-speaking Hayneses. I was quite torn between the two; I had really enjoyed being with Hélène, and in a part of Quebec that was trying to find its identity. But then I also enjoyed living the high life with the Haynes family. They could not have been more generous towards me, and Cherry and I fell back into our easy friendship. Above all, I felt excited to be in a North American city that had a strong European influence.

Back in Boston, I found St Patrick's Day in 1968 disconcerting: the river had been dyed green, the beer was green, and the then-city-councillor and unsuccessful candidate for mayor of Boston, Louise Day Hicks, rode at the head of the parade in a vivid green outfit. I found myself literally retreating from this over-the-top manifestation of Irishness. In Dublin at the time, the St Patrick's Day parade was pretty low-key. To find in Boston this exuberant display with floats, marching bands and Irish dancing in the street seemed a bit incongruous. Since then, of course, the American way of celebrating St Patrick's Day has shown the Irish how to do it in Ireland; now our own festival is just as important an event in the country's calendar, and American-style parades, but with their own local character, are held up and down the country.

I had decided to do my thesis on European Community competition law under Professor Henry Steiner, but I received a tempting invitation for the time the thesis needed to be completed. An English friend of my mother's, Judy Armitage, who had a home near Ballina where she and her family came to enjoy the wild beauty of the West of Ireland, invited me to stay for a long weekend at her house in the Bahamas. Notwithstanding what I had said to the Griz early on in the year, it was not in my nature to duck out of work, and I worked intensively to complete my thesis in record time so that I could go, at Easter, to Nassau. Arriving from bitterly cold, rainy Boston, in darkness, to the sultry Nassau climate, I woke up late the next morning and, since nobody was around, wandered through a back garden and out onto the whitest beach with the clearest turquoise water I had ever seen. My soul soared with the staggering beauty of nature, and I thanked God, whatever God, for such perfection. It filled my whole being. I have had that feeling maybe a dozen times in my life at most. Years later I experienced it when I stood on the slipway at Lough Conn, bordering our home in County Mayo, shortly after my first visit to Rwanda in 1994. We had just bought the house. I stood there, alone, for some time, listening to the lake water lapping against the slipway and watching the sun glinting on a small sandy beach across the water. I felt an inner peace lifting the strain of the images I had seen in Rwanda, and thanked God for such perfection.

When I flew back from the Bahamas, Boston was in turmoil. Martin Luther King Jr. had just been assassinated in Memphis, Tennessee, and across the country black people had taken to the streets in protest. My cab driver would not take me through the black neighbourhoods, as they were too dangerous. We ended up making a ten-mile diversion right around Boston. For the next week, I, like millions of others, did nothing but watch the unforgettable images on television and cry. I was distraught. The civil rights movement had become very much part of my life in Harvard. It had a great influence on me, the fact that young people in the movement were prepared to take responsibility even at risk to their lives. The newspapers were full of the stories of black students in the South being turned away from classes and of police guarding the classrooms. The assassination of Martin Luther King Jr. was absolutely devastating; that this young, inspirational leader who espoused non-violence in the face of terrible oppression of his people, evoking the basic freedoms, could be cut down like this felt almost unbearable.

That year at Harvard was hugely important in my intellectual development and gave me a real sense that values are worth fighting for. I was struck by the confidence of my contemporaries and by their commitment to a sense of public service and civic duty; these young people were going to take action. In Ireland, by contrast, it seemed inappropriate for young people to voice their opinions or to be involved in decision making: that was for their elders and betters.

It was a very idealistic time in America, partly because of the Vietnam War. There was much talk of the morality of the draft, and of those who had gone to Canada to avoid it. We would sit around for hours discussing the war. I remember being in the Harvard Cinema watching Humphrey Bogart and Katharine Hepburn in *The African Queen*; everybody in the audience knew all the best lines and shouted them along with the film. Suddenly there was a banging on the doors and walls of the cinema, accompanied by yells of cele-bration. President Johnson had announced that he would not seek re-election because of his unpopularity over the Vietnam War. We came out into the street in our excitement to share the feeling that the war might now come to an end.

My father flew over to Boston for my graduation. My mother didn't like to fly, but my father (who had worked without holidays for most of his life) now loved to travel, and arrived in fantastic form. It was wonderful for me to have him there. He had various friends in the Boston area, people who had come to stay in Mount Falcon in Ballina for the fishing and who had been his patients and friends, and we were wined and dined in style during his stay. After the graduation ceremony, we went to New York for a holiday before returning to Ireland. To this day, I remember the shock of waking up in our hotel in New York, putting on the radio and hearing the words 'Kennedy' and 'assassination'. I thought it was a programme recalling the assassination of John F. Kennedy, but was aghast to find that Robert Kennedy had been murdered in California. So soon after the assassination of Martin Luther King Jr. there was something very chilling about a country where people who stood up for their beliefs were being killed. I rang my father, who was equally devastated. 'Mary', he said, 'I feel like getting off American soil.' It was time to come home.

I was glad to return to Ireland, even though I missed the high energy of Harvard. I spent some lazy summer months being spoiled by my parents in Mayo, then moved back into Westland Row with Henry and Adrian, who were still living there, and started to practise law as a barrister. For the first six months of practice, each barrister was required to do a period of apprenticeship, unpaid, with an established junior counsel. The language used is interesting: the apprentice is the 'devil' who works with a 'master'. The choice of who would be my master was a big issue, because I was one of only a tiny handful of women barristers at the Bar. The idea of having a young female barrister on the somewhat chauvinistic, conservative Western Circuit, where I would plead cases in the courts in the main towns of Galway and Mayo, seemed like a joke in itself as far as some of the men were concerned. Even more so that I had asked the shyest of men, John Willie O'Connor, a middle-aged bachelor, to be my master. But John Willie turned out to be a brilliant choice. He was an old-fashioned lawyer's lawyer – he wrote things on the back of an envelope – but he understood that law is about quality and integrity: the calling of a barrister is an independent profession to be proud of, and he instilled in me (and in Henry, who also devilled with him) a great sense of the profession.

Early on in my devilling, my mother decided to throw a drinks party for the barristers who had come to Ballina on circuit. She poured plenty of gin and tonic into everyone, and late in the evening, John Willie, as he sometimes did on these occasions, took out his false teeth and put them for safekeeping in the nearest ashtray. The evening ended happily and everyone stumbled back to where they were staying. The next day in court there was a panic: John Willie could not find his teeth! While I held the fort and somehow conveyed to the judge that my master was not quite yet in a position to address the court, Henry was despatched to Victoria House, where he found that the drawing room had been cleaned and the contents of the ashtrays tipped into the fireplace. So Henry got down on his knees and felt around in the cold ashes where he found the dentures. He cleaned them under the tap, ran back to the court house robing room and handed them to John Willie, who came into court, bowed and with a big smile and a, 'may it please your lordship', proceeded to open his case.★

There were some fine lawyers on the Western Circuit at that time, notably Tommy Conolly SC, one of Ireland's greatest constitutional lawyers. I soaked it all up, endured the teasing and enjoyed the camaraderie. From an early stage I decided I would try to specialise as much as possible in constitutional law, which meant I stayed more in Dublin where the High Court was sitting, passing up on the income of the 'running down' cases – the fairly straightforward actions for personal injuries that were the bread and butter of the profession. But I had decided that I wanted to teach law as well, and obtained a post as tutor at UCD in the law of torts, grateful for the opportunity.

My first class was in Earlsfort Terrace, in the vast hall used by the Literary and Historical Society, where the seats were steeply banked, amphitheatre-style. When I came in for my tutorial, my thirty or so students were all sitting high up in the back row. So I said, with what I hoped was Harvard panache, 'Oh do come nearer; take the seats at the front.' The students, of course, were only waiting for me to say that and they all stood up and jumped down over the seats and towards me in a great clatter. This made me smile and blush and lose my cool a little. But I managed to compose myself, stop grinning, and get on with the lecture. The students seemed young to me, straight out of school, aged seventeen or eighteen. I was soon able to engage them with stories of the civil rights movement in the United States, making comparison with the increasing tension 'up North'

★ John Willie O'Connor later became a Circuit Court judge. He was unfairly criticised in the media for appearing to go to sleep during a lengthy trial in the Central Criminal Court, but it turned out to have been caused by an illness that ended his life shortly afterwards. He was greatly mourned by the profession.

where Catholics were beginning their own anti-discrimination and civil rights campaigning.

The challenge of teaching is one I have enjoyed ever since. One of the great things I learned from it early on is how to keep going, to keep speaking fluently, even when an awkward question is to be fielded, using little tricks to give yourself time to compose an answer. Teaching benefited me greatly in my law practice. It was also a great way to keep abreast of academic research. Above all, I realised how much I loved the contact and exchanges with young minds open to ideas.

In early 1969 the opportunity arose to compete for the Reid Professorship of Constitutional and Criminal law at Trinity. The part-time professorship was open to those in practice at the Irish Bar, and it involved an examination and interview. This was exactly what I was looking for, particularly its constitutional element. It would allow me to engage with the latest research while developing the kind of teaching methods I had admired at Harvard. Its modest income would give me the freedom to develop my specialisation and be selective about the type of cases I would take on. I prepared thoroughly, took the exam, sat the interview and happily was appointed Reid Professor, starting a five-year term in October. Needless to say, my parents were brimming with pride. To celebrate, I bought my first car, a small burgundy Renault 4.

Early in the same year, I accepted an invitation from friends living in the south of England to visit them. Nick was working in London as a freelance cartoonist, and I decided to look him up. We had been in correspondence sporadically while I was at Harvard, but now that I was back in Dublin, I missed his presence in my life. We rekindled our friendship immediately. Nick introduced me to his friends, but mostly we spent the time alone together, talking and laughing. We were both attracted to, excited to be with, each other, and we began to contemplate a more serious relationship. Nick planned to return to Dublin, and talked about continuing as a cartoonist, but also, perhaps, going back to the law and qualifying as a solicitor.

Meanwhile, as well as the excitements of my personal life, there was the possibility to take my reform ideas to a new level. An election to the Dáil, the lower house of parliament, was held in the spring of 1969. This meant that elections to Seanad Eireann, the upper house, or Senate, would be held six weeks later. Six of the sixty Senate seats were university seats, Trinity* having three, directly elected by the alumni. Two of the three declared candidates, Dr Owen Sheehy-Skeffington and Professor William Jessop, were sitting senators, and John N. Ross had been a senator previously and was looking to regain his seat. All three were a generation older, and men.

It was an intriguing prospect, to run for the Senate as a young woman who, teaching law and practising law, would have a chance to influence the making of laws. I researched the powers and procedures (and limitations) of the Senate, the university constituency, and how to get one's name on the ballot. I discussed the matter with my friends, who stood ready to support me. Nick, still in London, was the most encouraging, full of strategy and ideas. But my family had never been involved in politics, and I went down to Ballina to sound them out. I approached my father while he was in the garden digging his beloved spuds. 'I was thinking of standing for the Senate,' I told him. Long pause. My father looked at me as if I were mad. Seeing that I was in earnest, he suggested we drive out that evening to discuss it with Canon Luce, a retired Anglican clergyman who had been a leading academic for many years at Trinity.

AA Luce was an ardent fisherman, living in a modest bungalow beside Lough Conn, and my father was his doctor as well as his friend. He greeted us warmly and, contrary to my father's hope that he would douse the idea with cold water, was immediately enthusiastic. He had always been impressed, he said as he poured my father a glass of sherry, by my grasp of the Trinity Calendar, an annual compendium of the complex college statutes, appointment of elected officials, and various committees, which very few people read. He then spoke at length about the more obscure provisos of this document, which, in turn, impressed my father. I kept quiet, as

---

* Strictly speaking, Dublin University, of which Trinity is the only constituent college.

I had only the vaguest idea of the contents of the calendar and could not, in any case, see the connection between knowledge of its arcane contents and the skills required to embark on a political contest. Yet, if this was what it took to sway my father towards supporting my political ambitions, so be it.

From that moment my parents plunged themselves into the campaign, as did my brothers, and our friends and relations, with the excitement and enthusiasm of total amateurs.

My own confidence to run stemmed from the experience gained during that year at Harvard Law School. Nick teasingly called it my Harvard humility, but I had learned from my contemporaries that there were no boundaries to what we could seek to achieve. I had strong views on the need for change in Irish law and society, and this seemed a good platform from which to express them and to work towards implementing change. I could run for the Senate as an independent candidate, unlike contesting for a Dáil seat, where political party membership was almost obligatory. I wanted to have an independent voice and was not tempted to join a party.

I announced that I was running as a young person, a lawyer, a liberal, and a woman. Professor Bueno McKenna, then Regius Professor of Law, proposed me, and I was seconded by Nick's father, Howard W. Robinson, who had attended my auditor's address and been strongly supportive of me since then.

The electorate was comprised of approximately eight thousand graduates of Trinity. The team I assembled wrote letters to everybody we could think of, and asked everybody we knew in the Trinity establishment to write their letters in support of me. My family and friends worked morning, noon, and night targeting doctors, lawyers, friends of friends. I joked that I had sewn up the 'nun's vote' as my aunt Ivy, in India, looked to be the only one on the electoral register. Adrian had been president of the students' union in his final year, and a young woman from Millstreet, County Cork, Ann Lane, who had worked as a secretary with the union, volunteered to help us, especially with preparation of the many letters and the election literature. (Ann later became my secretary and personal assistant and worked with me throughout my legal career and during my term as president.) Nick, from London, helped me compose my election manifesto, and canvassed Trinity graduates there.

An impressive list of supporters had to be rounded up. Each candidate needed a proposer, a seconder, and five nominators. Ernest Walton, a Nobel laureate for his work in splitting the atom, was one I was anxious to enlist. I had sent him a draft of my election manifesto and visited his dusty office at the science end of Trinity. He sat me down and asked me a lot of questions before telling me he'd be delighted to nominate me. That was a big validation, and I was deeply appreciative.

My platform was essentially the need for reform in the areas I had earlier highlighted in my inaugural address as Law Society auditor: issues of law and morality in Ireland, the dubious special relationship between the state and the Catholic Church, the constitutional prohibition on divorce, the ban on the use of contraceptives, and the criminalisation of homosexuality.

The election count was held in the Exam Hall. Nick was there to do a tally, with my brothers and many of my helpers and supporters. The atmosphere was tense, so I took myself for a walk up Grafton Street and around Stephen's Green to clear my head and reflect. I knew we had fought a good campaign, that people had responded to it and that the Trinity electorate, mostly Protestant, was generally liberal and receptive to my ideas. But we were still up against the odds.

When I returned to the count, it was clear that I was going to win the third seat. My supporters were celebrating – Nick put his arms around me and gave me a huge hug – and I remember feeling humbled, relieved and elated. For the press, my election *was* an unexpected result – I had ousted former senator John Ross to win the seat, I was a woman, and I had just turned twenty-five – and both the *Irish Independent* and *Irish Times* had a prominent piece about it the next morning.

Winning the campaign had been an all-absorbing adventure, but now, here I was, an elected senator with all its accompanying responsibilities and challenges. What was I going to do now? Where was I going to start? God, I was so naive.

5

Ploughing My Own Furrow

As I began to climb the ladder, I had to cope with the different vocabularies used to describe similar qualities in men (confident, take-charge, committed) and women (bossy, aggressive, emotional).

Madeleine Albright, *Madam Secretary*

JUST BEFORE the beginning of the new term, in the autumn of 1969, I went into Leinster House, seat of the Oireachtas (both Houses of the Irish parliament), to deal with formalities such as signing the Senate register. There I met Senator Kit Ahern, a Fianna Fáil senator of long standing, who, welcoming me, looked me up and down and then told me, with some emphasis, that lady senators wore hats in the Senate. I only had one hat, a small one. Anxious to observe the rules, I duly wore it for the first day of the Senate term.

I found my way to the chamber, with its sixty seats of blue leather arranged in a semicircle around the ornate seat of the Cathaoirleach, or chair of the Senate. The first business was to elect a new Cathaoirleach. Senator Michael Yeats, son of the poet WB Yeats (who had also been a senator in his time), was put forward. As a member of the government party, Fianna Fáil, he was opposed by Owen Sheehy-Skeffington as not being sufficiently independent. I supported Owen's stance and, as the procedures required, stood up beside him – a little self-conscious in my hat – the only other senator to object to the proposed chair. My first official action as a senator was thus physical rather than oral. An inauspicious start, but I would find my (often awkward) voice soon enough.

On discovering that the wearing of hats was voluntary, not a requirement, I ditched mine there and then. (Having been photographed on the first day, it haunted me in the newspapers for months.)

I experienced a similar issue in the Law Library, the area in the Four Courts building in Dublin where barristers – all practising as sole traders out of a central library, as distinct from the chambers system used in the United Kingdom – congregated and worked when not in court. I liked to wear a smart navy pinstriped trouser suit in the Law Library, and under my black gown when in court. I had been absent from the Law Library for a few days, and on my return I found a notice pinned up in the ladies' robing room, where we donned our gowns, wigs, and tabs before going to court. It read, 'At a meeting of the ladies of the Bar it was decided that ladies would not wear trousers.' I took a pen out of my handbag and amended that notice to read, 'At a meeting of <u>some of</u> the ladies of the Bar it was decided that <u>some of</u> the ladies would not wear trousers.' I heard no more.

My life had become busier, and I was learning to compartmentalise my activities. I was still living in Westland Row with Henry and Adrian, and seeing Nick more regularly since he had moved back to Dublin, becoming a cartoonist with the *Irish Times*, while still talking about possibly going back to the law. I was teaching law in Trinity as Reid Professor, practising law mainly in the Dublin courts, and about to engage in law-making as a member of the Senate. I relished the ability to carry out these different activities, and worked hard so that I could manage them all. At the same time, I had resumed my friendship with Eavan, and even thought of taking up acting in the Focus Theatre, until my shortcomings on the stage were cruelly exposed. I decided to confine my performance to the theatre of the Irish courts.

The Senate was an impressive body for a young member like me. Although its legislative powers, as the upper house, were limited, I felt privileged to participate in law-making affecting the whole country. I admired greatly, among others, the contributions of Alexis Fitzgerald, Eoin Ryan, Jim Dooge, John Horgan, and later Conor Cruise O'Brien and Justin Keating; their speeches were thoughtful, sometimes scholarly. Even Michael Yeats, whom I had opposed as Cathaoirleach, was a fine chair and a committed and articulate senator. These were among my role models as I honed my skills in

addressing the issues being debated, in holding government ministers to account for what they had done, and in scrutinising what
they proposed.

In particular I regarded Owen Sheehy-Skeffington as a mentor. I
admired his intellect and independence of mind, and his commitment to opposing some of the terrible planning decisions that were
being taken at the time.

One campaign Owen was involved in, run mainly by students,
was to save from destruction a number of Georgian houses near St
Stephen's Green in Dublin's city centre. It was known as the 'Battle
for Hume Street'. Nick and I had been involved to a small extent in
this project with Owen and his fellow campaigners, among them
Niall Montgomery (my brother Adrian's future father-in-law, a
dapper, cultivated architect and intellectual). Nevertheless, I was
surprised to be woken by a telephone call one morning in June 1970
from a student leader of the campaign, asking if I would give a speech
at a public meeting scheduled for that afternoon. I said that while I'd
be happy to do so, they should really ask Senator Sheehy-Skeffington.
It was then that I learned the sad news that I was being asked because
he had died suddenly the night before, aged sixty-one.

I gave an emotional speech to the campaigners that afternoon,
recalling my mentor and friend and taking on his cause. Nick and I
became more involved with the campaign – a battle that was sadly
lost but that marked the beginning of a greater consciousness on the
part of Dubliners of their heritage and the need to preserve it. For
Nick it sowed the seeds of a personal campaign in conservation
work that would become a passionate cause.

In the Senate, I was determined to carry through on my election
manifesto, and to implement law reforms. My first objective was to
amend the law on family planning. The sale, supply, and distribution
of contraceptives in Ireland were criminal offences. It was legal to
use contraceptives, but not to buy or sell them. To me, changing this
was incredibly obvious; the law, in Shakespeare's terms, was an ass.
Under some kind of loophole, married women could be prescribed
the contraceptive pill by a doctor if they suffered from menstrual
cycle irregularities. We used to joke that it must be something to do

with the Irish weather, that such a sizeable proportion of Irish married women appeared to suffer from cycle regulation problems.

In October 1970, I announced at a public meeting of a group made up of medical practitioners and concerned citizens called the Irish Family Planning Association that I intended to introduce a private member's bill in the Senate to repeal the Criminal Law Amendment Act of 1935, the act that banned contraceptives. In doing so, I sparked off a heated debate in the media about the issue. The liberal *Irish Times* endorsed my position in an editorial, but the general Catholic stance was startling in its vehement opposition to this move. I had expected some criticism, but the scale and intensity of it shook me. 'Such a measure', proclaimed Archbishop McQuaid (in a letter to be read out loud at Sunday Mass in every church in his archdiocese), 'would be, and would remain, a curse upon our country.' The *Sunday Press*, a newspaper sympathetic to Fianna Fáil, printed the headline, 'A CURSE UPON OUR COUNTRY' in two-inch-high letters on its front page.

The year 1970 was a significant one in my personal life as well. Nick and I, having rekindled our friendship, which had deepened into a romantic relationship that drew me out of myself in an incredible way, had now committed to spend the rest of our lives together. He was the one for me. The trouble was, my parents were very much opposed to the match. In a way, perhaps, no one would have been good enough for me; certainly they had imagined an ideal husband for their beloved only daughter – he should be a professional, a doctor or lawyer, well established, moving in the 'right' social circles, and a paid-up Catholic – and Nick, budding cartoonist and Protestant, did not fit these criteria!

Nick's family, on the other hand, was entirely supportive. Nick's father, Howard Robinson, whom the family called 'the Boss', was a man of great principle and learning, had been president of the Institute of Chartered Accountants in Ireland, and as a volunteer worked hard and skilfully to safeguard the finances of the Church of Ireland. We had developed a strong friendship, seeking each other out for serious conversation.

Each of Nick's brothers was engaged in his own particular area of expertise. Michael, an actuary, was, with the encouragement of his wife Bobby, building in his spare time a forty-foot catamaran of his own design in his back garden. Peter had trained in the École Hotelière in Lausanne, and would shortly open his own Michelin-starred restaurant, Armstrong's Barn, in Wicklow, with his wife Christine. Andrew and his wife Jenny were classical musicians and teachers, and he also made musical instruments (violins and viola da gambas) at his home. The Robinson family background was a contrast – but a pleasing one – to the legal, medical, and sporting world in which I had grown up. I felt very comfortable with the wider Robinson clan, including the Boss's two sisters, Phyllis and Audrey.

My parents asked me to take a six-month break from my relationship with Nick. Nick and I discussed the idea. We were confident that our feelings weren't going to change, but if this would prove in some way to my parents that we were serious about our commitment to each other – if they would then accept Nick – then maybe it would be worth it; so we agreed. Durnig the six-month moratorium, we did not correspond, and met only once, and that was accidental, when we were both invited to a friend's drinks party.

It was tough going. I was finding my feet in the Senate and would dearly have loved to talk to Nick about all the goings-on. But the time passed, and as soon as the six months were up, we got back together again. My parents were furious. My father was particularly angry and thought I was exercising poor judgment in thinking that Nick would be an appropriate husband. They let me know in no uncertain terms that they would not support the match and would take no part in it. It pained me to see how upset they were but this time my mind was made up, I couldn't be separated from Nick anymore. Once again, as with my lonely path towards finding my own spirituality, I decided that there was no point in arguing further, but that I should just go ahead with Nick in our plans to get married. This was my life, after all. Nick understood that my parents were reacting this way because they loved me so much. He assured me that they would eventually come around and that he would not bear any grudge.

We needed permission for me, as a Catholic, to marry a Protestant, so we went to see Monsignor Sheehy in the Archbishop's Palace in Dublin. He left Nick in an unwelcoming, cold, windowless room and sat me down in his study and interrogated me. Why did I want to marry Nick? Would the children be brought up Catholic? Was I still loyal to the Catholic faith of my childhood? I was enraged by the whole process and infuriated further when I felt tears coming. But the sight of my tears discomfited the monsignor, and he assured me that I would have the permission. Red-eyed, I found Nick to tell him the news.

I met Eavan – recently married to the writer Kevin Casey – in a café on Grafton Street and told her about our wedding plans, but that there'd been a bit of family weather. She understood me, understood how that could happen. She listened and then said, 'Why don't you get married from our new house?' I readily agreed. Then she began to laugh, telling me that this was an incredibly optimistic idea since they didn't yet have the keys!

We planned to marry early in the morning in the church at Dublin Airport so that after a wedding breakfast we could catch the morning flight to Paris to begin our honeymoon. Mary O'Donnell, the couturier and a friend of my mother's, made my dress, and a first cousin of mine, also Mary O'Donnell, agreed to be my bridesmaid. She would be the only member of my family to attend. A couple of days before the wedding, I was still living in Westland Row and met Henry there, who said he wanted to come. But my oldest brother, Ollie, stopped him short, enforcing the family line. In truth, I didn't encourage any of them to come. I didn't want the family to split over this. If it was just me, it would make the mending of the relationship easier.

I had originally hoped that my good friend Father Enda McDonagh would marry us, but anticipating that there would be parish difficulties he suggested we ask Father Jack Kelly. Enda knew I had been friendly with Jack as a student, and Nick and I agreed that he was a great choice. To reflect both Christian traditions, we invited the dean of Christchurch Cathedral, Tom Salmon, a Robinson family friend, to take part in the wedding.

When Eavan and Kevin moved into their new home in Dundrum, I moved in with them. It was exceedingly Bohemian; everything disorganised, packing cases everywhere. It was lovely.

We had a rehearsal at the Dublin Airport Church the evening before the wedding. Having come from a reception given in his honour, Jack was slightly merry and kept making witty asides as he talked us through the service. The laughter was therapeutic and dissolved some of the tension that had been building. After the rehearsal, we dropped Jack back into town, and he came into Lincoln's Inn with us for a last drink. He huddled the two of us in together with him and raised his glass as if in a toast: 'Who do these Bourkes of Ballina think they are anyway?'

As Nick and I were saying goodbye, I asked him, 'Well, how do you feel? Tomorrow your freedom ends!' He smiled. 'When I marry you my freedom begins.'

Very early on the morning of 12 December 1970, Eavan woke me with a cup of coffee. My wedding dress was hanging on the back of the door and had a slight snag on it that was bothering me. It was no big deal, but I asked Eavan if she had a needle and thread. She looked at me, one eyebrow raised. 'I can't even find Kevin in here! What do you mean do I have a needle and thread?' We laughed, and Eavan helped me get dressed. We had breakfast. Then Judge John Kenny, a friend and mentor of mine and a close friend of Nick's father who was to bring me to the church and give me away, arrived at Eavan's in a Rolls-Royce. I got in and we drove to Dublin Airport. On the way, John spoke about anything and everything; he talked about the law, he asked me about the Reid professorship, he talked about the Senate, anything but personal matters. I appreciated that. For all my certainty and obstinacy, putting a brave face on it, I had spent one of the loneliest nights of my life, and I think if John had said anything relating to my family I would have burst into tears.

The wedding Mass, with its joint blessing by Father Kelly and Dean Salmon, was conducted impeccably – Father Kelly in his homily wished for us, 'May you grow old in the company of your friends' – and was followed by a lovely wedding breakfast attended by friends and all of Nick's family. Then we took our flight to Paris, a married couple.

News of our wedding had belatedly leaked out to the media, and some journalists and photographers arrived shortly after our departure. Nick's friends at the *Irish Times*, where his political cartoons were published, had loyally respected his request for privacy. We were relieved by this; sensitive to the circumstances of defying my parents to marry Nick, I did not want them to come across a colour piece about it in the newspapers. We never showed my parents any wedding photograph – I have always felt a twinge of pain about that – although those of my brothers still adorn the walls of the smoke room in Victoria House.

It was a beautiful day when we arrived in Paris, bright wintry sunshine and blue sky. In the morning we walked the banks of the Seine and took photographs of each other. We retreated to our hotel at the cheap end of the Île St-Louis for the afternoon, and that evening went for dinner together at a glamorous restaurant, Maxims. We were amused to note that we were easily the youngest couple there; well, to be more accurate, I was not necessarily the youngest woman there, but Nick, at twenty-four, was by far the youngest man! We savoured the food and wine, and then called for the bill. To our mild horror, we discovered that we had not brought quite enough cash to cover the bill; we were about seven francs short. We explained this somewhat red-faced to the waiter, offered in a mock-serious tone that Nick would do the washing up, and then played our irresistible trump card: 'We're on our honeymoon.' The waiter went away and came back beaming. 'That is no problem, monsieur.' When, however, we went to collect our coats in the cloakroom, and had not a centime to offer the lady looking for her *pourboire*, she emitted a string of inelegant French and chased us off the premises. Laughing, arm in arm, we took the long walk back to our hotel.

Next day we flew to Tenerife to stay for three weeks in a bunga-low owned by Nick's father, near the village of Candelaria. There Nick learned to appreciate the only dish I could cook, 'a delicious concoction' he assured me, involving eggs, bacon, beans and left-overs baked together in the same pan, which we christened 'huevos Maria'. We began to learn Spanish with some taped lessons, laughing as Nick mimicked the high female voice on the tape.

Returning from our honeymoon we planned to live in a house in Wellington Place that we had bought from Nick's brother Peter. Peter hadn't yet moved to Wicklow, where he and Christine would open their restaurant, Armstrong's Barn, so for the first months of our married life we lived in my father-in-law's mews house, at the bottom of the garden, surrounded by pieces from his art collection.

I missed my parents terribly. Since our wedding day my first thought on waking every morning had been to reconcile, and several times I almost picked up the telephone to call them, but my pride stopped me. My great-aunt, Mother Aquinas, wrote to me to say that my parents were very upset and what was all this, why wouldn't I contact them? I wrote back that they were the ones who had boycotted my wedding; it was up to them to make the first move. Aquinas dismissed this as nonsense. Still, I couldn't bring myself to make the telephone call, much as I was longing to make peace. Two months passed.

At the end of February I was invited to take part in a live television programme in Belfast with Barbara Castle, the formidable English Labour politician, and then member of cabinet. I readily accepted, and Nick and I drove up. There was still some snow on the ground, and as we drove we hit a patch of black ice and the car spun around and gently slid into a ditch. We were a bit shaken but not injured, and shortly afterwards a tractor drove past and the driver spotted us. She had a rope and was able to pull us out without much difficulty, and we drove on to Belfast to do the television programme. Driving back home to Dublin afterwards, I kept thinking, what if we had been badly injured, or killed? What would that have done to my parents? As soon as we got home I rang my mother in Westland Row. I knew she was visiting Henry and Barbara, who had married the previous September and were living there. She responded warmly and agreed to meet for a cup of coffee in town the next day. As soon as we met we hugged, overjoyed to see each other. We managed to skirt around the whole subject of the wedding and the falling-out; and before we parted she asked me to bring Nick down to Ballina for the weekend.

Nick and I drove down, and when we reached the nearby town of Swinford, I asked him to stop so we could buy a bottle of champagne, as we would be celebrating. This set the tone. We produced the champagne as soon as we arrived, it was put on ice, and we all sat in the drawing room in Victoria House and drank it together. My parents were delighted to see us, my father beaming. We never ever talked about the rift. It had been a moment, but I didn't want it to define the loving relationship I had with my parents. Thanks to Nick's generosity of spirit, my parents and he got on well from then on. He and my mother forged a close bond – she liked to refer to him as her favourite son-in-law! – finding that they had much in common, especially when it came to doing up our new home in Wellington Place. My parents, of course, had an unshakeable respect for the sanctity of marriage, so that once I was married, in law and before God, they would never have sought to undermine it in any way.

I had built up a reputation, and endured some notoriety, as Senator Mary Bourke. At that time, women professionals were beginning to keep their own surnames, but I decided that because of my parents' opposition to my marriage, one of the ways I was going to seal it was to take Nick's name. I teased Nick that I was making a huge political sacrifice because my surname would of course appear alphabetically, right down towards the end of any ballot paper, which statistically would put me at a disadvantage, albeit slight.

In early 1971 I carried out my commitment to introduce a bill in the Senate to amend the law prohibiting the sale, supply, and distribution of contraceptives in Ireland. Any member of the Senate or Dáil could introduce a bill once it was supported by two other members of the House. John Horgan and Trevor West agreed to support my bill. John was an *Irish Times* correspondent who was also in his first term as senator, elected from the National University of Ireland; he became a close friend and ally in a number of subsequent private members' bills. Trevor had been my election agent and was elected in late 1970 at the Trinity by-election following Owen Sheehy-Skeffington's untimely death. It was a simple, technical bill, a few lines on one page, repealing the prohibition contained in the 1935 Criminal Law Amendment Act.

Normally, a private member's bill, once submitted, would automatically be approved by the Senate for a first reading. This simply meant that the bill was printed as a Seanad Éireann bill, published and circulated to the members of the Oireachtas. It was, perhaps, unheard of for a vote or any controversy to arise in respect of a bill before this official publication, the first stage. But that is exactly what happened. The Senate leader, Tommy Mullins, a member of the government party, Fianna Fáil, dictated the day-to-day order of business of the Senate. For months, he skipped over my bill, citing pressure of government business. I argued that all we sought was passage of the first stage so that the bill could be published and debated in the House, that to deny this was effectively an interference with the democratic process, a form of censorship.

Meanwhile, although the bill had not been published officially in parliament, its contents were known, and the debate continued to rage in the media, with many newspaper column inches and radio and television time given over to it. One of the points we tried to make in the Senate was that everyone in the country seemed to be debating the issue except the legislators. There was an Irish tendency to dodge sexual and moral issues by upholding Catholic principles in the law, while in practice doing otherwise. I was offending this fudge by wanting the law to reflect the need for openness and diversity in Irish society. Access to contraception, I argued, was a matter of private morality. But even the leader of the Labour Party at the time, Brendan Corish, drew me aside to express his concerns about a bill of that kind because it offended the Roman Catholic ethos of the country. It alarmed me that a good and decent political leader would either not make the distinction or bow to such pressure.

The hierarchy of the Catholic Church rose up against the bill, condemning it from the pulpit and denouncing me, personally. This did not cause me grief; I already felt distanced from that authoritarian orthodoxy that I believed was harmful to women's health. Nor did it undermine my personal spiritual path; I was seeking to follow the gospel of Jesus in a more abstract fashion. But then I was denounced from the pulpit by the bishop in Ballina Cathedral, and learned from my brother Aubrey (who was supportive of what I was doing but wanted to alert me that it was taking its toll) that my parents had left the cathedral in terrible distress. I became increasingly concerned about the way in which the controversy was very publicly affecting them.

The degree of fury and venom to which I was subjected was sobering and weighed on me, but the real pressure came from the effect this criticism was having on my parents. The pain it was causing them as devout Catholics was evident. My father looked older and more lined, and my mother had lost her sparkle. They did not try to stop me from pursuing the bill. We didn't ever directly discuss it, at that time or later. That would have caused them further pain, as I knew their Catholic beliefs would outweigh the arguments I would bring to bear. Having been reconciled with them so recently, I longed to ease their hurt, but I was nevertheless determined to persist with the bill and get it through the Senate. I did not see any other way but to press forward.

Naïvely, I decided I would try to see Cardinal William Conway, the Catholic Primate of All Ireland, and explain to him my reasoning so that he might somehow send a message to Bishop McDonnell in Ballina, who could then have a word with my parents, offer them some comfort. I made contact and was invited to meet His Eminence a few days later, in Dundalk.

Nick and I drove up together, refining my arguments as to why it was important to make the legislative change and bring Irish law more into conformity with the practical reality of Irish life. I was full of hope that a burden would be lifted. We arrived. As planned, Nick remained in the car, while I was shown into a waiting room. After some time I was told the cardinal would see me.

I had seen photographs and television images of Cardinal Conway, but we had never met. He rose when I came into the room, a large, imposing presence who greeted me with a northern accent and no smile or warmth. He was dressed in a long black cassock, with his cardinal's scarlet zucchetto. I shook his hand, knowing that he might have hoped I would bow and kiss his ring. I had an immediate sense of being in the presence of a prince of the Church who felt he had graciously condescended to grant me a brief audience. Reflecting afterwards, I think it was because I was so full of hope and expectation that this meeting would help change the public perception of what I was trying to do, that I felt so disheartened. The cardinal was not in the slightest bit receptive to the points I had rehearsed, and responded with long references to canon law and the Catholic ethos of Ireland. There was no moment when we shared common ground. I explained about my concern for my parents and asked him to contact Bishop McDonnell. But he would give me no comfort. 'They are good people' was all he would say. 'I will pray for them.' I sensed a terrible coldness towards me, this Trinity liberal who was trying to undermine the Church's authority, damage his power base. There were no tears this time. I was disappointed, but I left the cardinal with a determination that I would not be bullied further.

At home and in my post box in Leinster House, I received hate mail. Whether it was coordinated, I did not know, but it came in quantity every time a new development happened in relation to the bill. 'You must be the Devil incarnate to do this!' 'May you burn in Hell, from a Catholic Priest.' Some were typed, some handwritten. They included (as crude imitation condoms), cut-up fingers of garden gloves. This reaction affected me very much. I was used to being admired and supported and it was a whole new experience for me not only to be criticised but to be actively hated by people, strangers. I hid my feelings, but I remember walking down Grafton Street in Dublin's city centre and feeling that people were looking at me. I was walking defensively, waiting for someone to jump out and tell me how much they hated me, that they wished I would burn in hell. That is when Nick decided that these letters were getting to me, and we should burn them. Shortly afterwards, we regretted destroying them, as they would have made an important archival record of the extent to which some people felt threatened by the implications of the bill.

Then a rumour began circulating that Nick was related to the owners of a successful pharmaceutical chain, Hayes Conyngham and Robinson, and that I would be paid a penny for every pill and condom sold! To offer such a grubby commercial motive was effective. It didn't matter that it was utterly spurious.

In the end, the bill never even achieved a first reading. John Horgan, Trevor West, and I unsuccessfully attempted, on a number of occasions, to have the Senate's order of business amended to allow the bill to pass the first stage and be published. Out of the blue, one lunchtime near the end of the summer term in 1971 – fortunately, I was in Leinster House at the time – I was informed that the government had taken the unprecedented position that it was opposing a first reading, and the bill was to be moved and opposed first thing that afternoon. I had to rush up to the chamber and speak to it straight away. I was furious to be taken by surprise like this and given no notice or opportunity to prepare properly what I would say. It was, at the very minimum, a gross discourtesy. I confined myself to making what I described, at the time, as 'a rather angry statement'. We were thus outmanoeuvred, and the bill was killed.

I was furious at the way our initiative, supported by many people in Ireland despite the controversy, had been thwarted without even a debate on its merits in the Senate. What were political leaders afraid of? Why could they not at least discuss the merits of giving a choice about family planning? Naively, I had believed that it would be relatively straightforward to introduce an amendment to the law and get it accepted. I had focused solely on the legal issue, not appreciating the political realities. It was a hard lesson to learn that there was not only political opposition but also a cultural lack of willingness in Irish society to talk about sexual matters. Even mentioning contraception was more than most people wanted to do. Professor Enda McDonagh captured the mood of the time in a recent essay titled 'Culture and Religion in Ireland, 1960–2010'.

A censorious approach was taken to the life of the senses generally; human existence was referred to as 'mourning and weeping in this vale of tears', as a favourite hymn put it. Sexual repression, as it is now called, was strongly associated in the public mind with religious dominance, which in turn tended to be blamed for strict state censorship of books and films. The cultural life of the nation, both in its specialist ('high') and popular forms, was impoverished. And all this was compounded by a general anti-intellectualism, widespread poverty and increasing emigration . . .

The poorly understood distinction between culture and religion, their intermingling at so many levels and in so many areas of life, from the artistic and intellectual to the political and economic, obscured what in neighbouring countries would be sharply differentiated as properly secular and properly religious.

I was not going to give up, and with my co-sponsors, and through what might be characterised as obstinacy, I kept the issue alive. In the next parliament, under a Fine Gael–Labour coalition government, in 1973, we succeeded in getting a first reading for an expanded bill, now called the Health (Family Planning) Bill, which proposed to deal with contraceptives as a health matter rather than a criminal law matter. In 1974, a court case, *McGee v. Attorney General*, in which I took no part (though I am sometimes, mistakenly, given credit), found that the seizure by the Irish customs authority of contraceptives sent to Mrs McGee from Britain was an unlawful interference with her constitutional right to privacy. This was one of a number of landmark Supreme Court judgments of the time that identified 'unenumerated' personal rights as being implicit in the wording of the Constitution.

The pressure worked in that it led to the coalition government introducing its own modest, conservative Family Planning Bill in the Dáil, though what happened next was like something out of a farce. The government, having introduced the bill, then unexpectedly announced a free vote, which meant that individual deputies could vote according to their conscience and did not have to follow the party line. The Taoiseach (Prime Minister), Liam Cosgrave, and some cabinet members, in an astonishing move, acted as conservative Catholics rather than government ministers and defected to join the opposition side and vote against their own bill, to the shock and embarrassment of colleagues, ensuring the bill's defeat. A subsequent Fianna Fáil government did eventually, in 1979, introduce a limited, highly regulated form of legal contraception in Ireland in what Charles Haughey, then Minister for Health, described as 'an Irish solution to an Irish problem'.

With hindsight, I fault myself for being so judgmental, thinking that one could simply propose the amendment of a law that reflected such a fundamental belief when a desire for change wasn't coming broadly from within the country. The revelation of the depth of the trauma for Ireland in beginning to talk about things that were anathema and for so long taboo was a huge lesson. Though I still believe – on *that* issue, in *that* Ireland – there was no way to do it other than to introduce legislation and start a debate. Later it would make me more patient and sensitive in the context of human rights issues such as early child marriage or female genital mutilation, understanding the need to work on education and to discuss an issue extensively before proposing a change in the law. Even if a law is changed, real change, in practice, has to come from within communities themselves. I learned not to underestimate the compelling hold of 'traditional practice'. I learned to describe these issues as 'harmful traditional practices' rather than as 'culture', which would give them an unwarranted importance.

I also learned another hard lesson. If you speak out for something you believe in, and take on these harmful traditional practices, you have to be willing to pay a personal price. Included in the price I paid in this episode was the deep distress suffered by my parents on hearing me being denounced from the pulpit of their cathedral.

I often draw on this experience and encourage young people to understand that if you really believe in something and are prepared to pay the price, you are then free to speak truth to power.

6

Looking Beyond Our Borders

We are part of the community of Europe, and we must do our duty as such.
William Gladstone, speech, 10 April 1888

I TOOK the unusual move for a barrister, at the time, of establishing an office outside the Law Library. It was part of my strategy of being able to compartmentalise my various working engagements, and have time for the domestic part of my life. I never considered myself a full-time barrister; that was always a part-time job, as was teaching, fitting in around my other work, including as senator.

My office was located on the ground floor of No 27 Merrion Square, an elegantly proportioned Georgian house in Dublin's city centre that had been converted into offices. Adrian's father-in-law, Niall Montgomery, whom I had got to know well through our mutual involvement in the Hume Street campaign, conducted his architectural practice from the top floor of No 27. Ann Lane joined me as my full-time secretary and was quick to adapt to the work. We became good friends and worked well together, she understanding how I operated, what I needed from her, and how to deal with my clients and colleagues, occasionally acting as the dragon at the gate when I did not want to be disturbed. Although she was a petite woman, she had an earthy sense of humour, was an intrepid mountain climber, and rode a powerful Honda motorcycle. Occasionally I would don her spare helmet, jump up on the pillion, and she would give me a lift to the Four Courts or to Leinster House.

During this period, I became more and more convinced that a major part of my energies should be devoted to the cause of Ireland's membership in the European Community. My academic work, starting with my LLM thesis in Harvard, and leading on to teaching a course in Trinity on European Community law, convinced me that

Ireland's best future was in Europe. This led me to join the European Movement, one of whose primary objectives was Ireland's entry to the European Economic Community (EEC), as it was then known, now the European Union (EU). Through the movement, I deepened professional relationships and personal friendships with some of the pioneers of Ireland's cause in Europe, notably Garret FitzGerald, Denis Corboy, Eoin Ryan, Tom O'Higgins, and Michael Sweetman.

Sometime in 1971, Denis Corboy, who was then head of the Information Office of the EEC in Dublin, and who encouraged me to focus not just on European law but also on the politics of the European Community, put my name forward to be the Irish member of the Vedel Committee on the enlargement of the EEC.★ Ireland, Denmark, and the United Kingdom were gearing up to join the EEC, then consisting of the six original member states: Belgium, France, Germany, Italy, Luxembourg, and the Netherlands.

Denis, full of charisma, was not affiliated to any political party, being a through-and-through Eurocrat; whenever he started to talk about the institutional or political developments in Europe, he became immensely serious and interesting. I had first met Michael Sweetman, equally committed to Ireland's joining the EEC, as a schoolgirl, his sister Mary being a direct contemporary of mine at Mount Anville. Reacquainting myself with him in the political sphere, I found in him an honesty of thought and a rounded political judgment reflected in his work with Declan Costello in drafting *Towards a Just Society*, the Fine Gael Party's impressive new manifesto. While my political inclination was more radical than his in terms of a reform agenda, we were in agreement over Ireland's need to join the EEC, and I spent some delightful time in his and Denis Corboy's company as we campaigned for the referendum on Ireland's membership.

When I met Garret FitzGerald he was already a well-known figure in Ireland. We got on very well from the beginning, and when

---

★   A committee of experts chaired by senior French parliamentarian Georges Vedel, that published its report on 25 March 1972.

I joined the Senate, he drew me aside and quietly asked if I would be interested in joining Fine Gael. Thanking him, I just as quietly declined the invitation, and we never broached the subject again. We later discovered that Garret and Nick shared a birthday, 9 February, Garret being exactly twenty years older than Nick. Garret's wife, Joan, and I decided to throw a birthday party for them, for their joint 'eightieth' birthday in 1976, and the families organised a joint celebration thereafter every five years.

I felt strongly that it was in Ireland's interests to become part of a wider European grouping of states. This would help us to stop seeing ourselves always reflected in the mirror of Britain, next door. Even then Ireland still had some of the feeling of inferiority of a post-colonial country. There had long been the sense that the English looked down on or tried to ignore Northern Ireland and didn't really think very highly of Ireland or Irish society, whereas the French were interested in our culture, the Germans were interested in our writers, the Italians were interested in the beauty of our landscape. I thought that joining the EEC and becoming part of a community of nine would help Ireland come out of the shadow of its more dominant neighbour and negotiate with the other member states in the context of EEC regulations and directives. I believed we could find common ground with France, for example, in agriculture, and with Britain on other aspects of legislation where we would have a common tradition, so that instead of a post-colonial semi-dependency, we would be more mature. I felt that it would actually reinforce our sense of identity, because we would take our place proudly as one of the members of the EEC. At the same time, I did respect the concern on the other side, which was most eloquently expressed by Justin Keating. Essentially he argued that Ireland had struggled for centuries and then for decades in the twentieth century to achieve independence, we had our independent constitution, we were no longer a member of the commonwealth, we stood on our own two feet. We needed to reinforce more our Irish heritage and language. This would become lost in a sort of homogenised Europe. Ours was a good intellectual engagement in the referendum debate, genuine disagreement, often between friends.

In the end I believe that joining the EEC did actually enhance Ireland's sense of identity, of culture, of political maturing. I sensed what would become true later: that there would be a huge opportunity to accelerate personal rights and freedoms, particularly equality for women, because of Article 119 of the EEC Treaty and the existing directives on equal pay for equal work. In a broader sense, I felt that the culture of Ireland would be appreciated by our European partners, and that we ourselves would see Ireland more as a modern European country and not as the former colony of the United Kingdom, with all the baggage of that relationship and perception.

The ongoing and increasingly violent Troubles in Northern Ireland were a black cloud looming over all of us, creating further tensions in this already difficult relationship with our nearest neighbour. Entering the EEC as an equal partner, albeit without the corresponding economic might, aligning our laws under this European regime, would, I hoped, also give us more credence with the Protestant tradition in Northern Ireland, as we would be part of the same wider European Community.

When two bombs went off in Dublin in December 1972, killing 2 and injuring 127, the Offences Against the State (Amendment) Bill was rushed through the Dáil and Senate (sitting late into the night), in what was clearly a knee-jerk reaction. I argued forcefully, but unsuccessfully, for an amendment that would give the bill the status of emergency legislation thus lasting for a period of only ninety days, after which it would have to be renewed. I thought this at least would give us time to debate the issues more fully and in a less emotive atmosphere. The legislation focused on prosecutions for membership of illegal organisations (i.e., the IRA) and controversially allowed the unsubstantiated evidence of a garda chief superintendent's belief that a person was a member of an illegal organisation to be admissible in a trial.

For me this was a first taste of how the legislature, reacting to a terrible 'terrorist' event, could put laws onto the books *restricting* rights, overturning what I believe is the constantly necessary stance against any *curtailment* of human rights even in the face of violent aggression. Soon after that I gave a lecture (which was subsequently published) critical of the Special Criminal Court, a court that heard criminal prosecutions without a jury, and was designed to deal with members of illegal organisations, but which was subsequently used (and at the time of this writing still is) to prosecute 'ordinary' serious offences such as organised crime (effectively evading trial by jury by the back door).

Working on the Vedel Committee allowed me to engage with foremost European academics and scholars to tease out the issues facing an enlarging Europe, to be at the birth, as it were, of a new legal order that would shape our own. This informed my thinking about political and legal reform towards equality, non-discrimination, and personal rights and freedoms. The Committee's work was to examine what impact three new member states (an increase from six to nine) would have on the institutions of the EEC, but also to try to increase its democratic legitimacy. So we proposed strengthening the powers of the European Parliament and making the work of the other institutions, the European Commission and the Council, more transparent, more responsible to the people they served. Some of our recommendations were taken on board, though the bolstering of the European Parliament came only years later.

It was an opportunity to help shape one of the key regional institutions at the time, particularly by strengthening its democratic base, which lacked accountability and was, even after our recommendations, not strong enough. The institutions of the European Union still exercise power without adequate transparency and responsiveness to the populations affected by their decisions.

During this work, I came up against another example of petty discrimination against women. I happened to be the only woman on the Vedel Committee, which met every couple of weeks in Brussels. The first time we met we were given forms to fill out, selecting a bank where we would open an account into which would be paid our stipend. In my case, I was asked if I was married, and when I replied that I was, I was told by the bank that my husband's written permission was required for me to open an account. Even though I came from a country where women were required to retire from their civil service jobs on marriage, where women were not permitted to sit on juries, where women had no entitlement to pay equal to that of their male counterparts (and the list went on), I was stunned by this requirement. Though Nick gleefully assured me he would give his written permission, I absolutely refused to entertain the idea that I would seek or provide them with it, and in the end we came to a compromise where I was dealt with by the bank as 'Ms' Robinson, as if an unmarried woman.

While I was enjoying teaching as Reid Professor of Constitutional and Criminal Law at Trinity, I suggested to the head of the law school, Professor Bueno McKenna, that, prior to Ireland's joining the EEC – which it would do at midnight on 31 December 1972 – Trinity should offer a course on European Community law. Professor McKenna said that if I was prepared to add it to my

My choice of title for the Reid Lecture I was to give in 1971 drew this response from Nick.

part-time teaching as Reid Professor, Trinity would give me a small further stipend. It became the first undergraduate course of its kind in the country, as far as I know. Teaching EEC law, I was keeping abreast of the way in which the EEC was developing, the treaty provisions, the approach to legislation, and was filing away useful directives on equality, non-discrimination, and the like for when they might come up during the course of my work. Denis Corboy's office gave me copies of the latest regulations and directives to share with my students.

Out of the blue, and at the proposal of Professor Max Kohnstamm, whom I had met on the European academic circuit, I joined the Trilateral Commission, a policy think tank involving high-powered membership from North America, Europe and Japan, whose purpose was to discuss issues of global concern. I became a member of the executive board when it formed in 1973. Nick came with me to these meetings, which offered us a steep learning curve and opportunities to travel widely and mix with leaders in industry and academia and with elected representatives. Other Irish members (in a body of more than two hundred, virtually all men) at the time included Garret FitzGerald and Ken Whitaker. We met in different regions on various issues of common interest – generally relating to trade, energy, and democracy – which usually led to reports being drawn up, co-authored by a member from each region. The main political objective at the time was to forge closer political ties between the United States, Europe, and Japan on issues of common interest.

At a commission meeting in Japan in 1975, I met Jimmy Carter, who was seeking to be the nominee of the Democratic Party for the US presidential election and was at the time a relatively little-known former governor of Georgia. I found him surprisingly approachable. It was said that through his work with the Trilateral Commission, he deepened his grasp of US foreign policy. Whether or not that was true, when he became president, he took more than a dozen US Trilateral members into his administration, including Vice President Walter Mondale, Secretary of State Warren Christopher, and National Security Adviser Zbigniew Brzezinski. Zbig had been very influential in the early days of the Trilateral Commission and I admired his vision of the world and the clarity of his thinking about finding a better balance between the United States and other countries.

When Nick and I were leaving that conference in Japan, we shared a hotel elevator with David Rockefeller, the American chair of the Trilateral Commission. He saw our suitcases and asked us if we were leaving, and we replied that yes, we were going home via Singapore. 'Oh, do give my very best regards to Lee Kuan Yew', he exclaimed, 'a remarkable young man, don't you think?' Lee Kuan Yew was Singapore's prime minister at the time, not particularly young and certainly not somebody we would know, but it was typical of a Trilateral Commission member to assume we would be paying such a courtesy call on our stopover.

While the commission's reports focused primarily on trade issues, I worked on a draft report with US executive member William Scranton and the Japanese chair, Takeshi Watanabe, in which we advocated a greater reaching out by the Trilateral Commission to the developing world, a closing of the gap between what we described then as the global North and South. That report was neither finalised nor published, the issue was largely ignored, and I gradually grew disillusioned, realising that I was fooling myself if I thought that this body, whose real interest involved interaction with the dynamic Asian market, was going to reach out and address the imbalances, unfairness, and trade barriers for developing countries. I resigned in January 1980, citing the commission's lack of real commitment to leadership and creativity in examining the issues of North-South relations, and expressing a concern about the increasing personal influence I felt was being exerted by Dr Henry Kissinger. On a personal level, I got on well with Kissinger; his forceful, brilliant intellect and curiosity for ideas made him fascinating company for me. But his vision of global relations and the role of the United States was much too conservative for my sense of what the Trilateral Commission should be doing.

Despite my resignation – George HW Bush resigned about the same time, believing the Commission to be too liberal – some journalists in Ireland persisted for years in attributing to me ongoing membership (and in one or two cases, probably still do), offering fantastic conspiracy theories as to how my career had been advanced by it.

Meanwhile, Nick's career had taken a new turn. Having qualified as a solicitor, he worked initially with Denis McDowell's firm, and then set up his own practice near Fitzwilliam Square. Over the years he would occasionally brief me as counsel, and working together intensified our bond.

While I was intently focused on my own thriving career, my personal life had undergone a major upheaval in 1972 when, to our intense joy, I became pregnant. Nick was enormously solicitous, so much so that I had to make it clear to him that pregnancy was a normal, healthy state, not an illness, and I didn't need to be so coddled. My parents were overjoyed of course; this was to be their first grandchild. My mother quickly signed me up to one of the most conservative Catholic gynaecologists in Dublin, a friend of hers. Somewhat reluctantly, I bowed to her choice. As my bump grew, some colleagues in the Law Library were slightly uncomfortable to find that I was continuing to work in my condition – there were so few women barristers they hadn't yet encountered a pregnant colleague – but I worked almost up to my due date, finding that it suited me and my psyche better, the idea of resting up at home being anathema.

In June of that year, 1972, we heard the very distressing news that a plane taking off from Heathrow to Brussels had crashed near the town of Staines, in England, killing all on board, including a delegation of Irish business leaders. Among them was my friend, the brilliant Michael Sweetman, aged thirty-eight. I had met his wife, Barbara, only once or twice, and did not really know her in her own right then, only as Michael's spouse. Michael had talked of her and their six children with great affection on our trips abroad with the European Movement.

Nick and I called to see Barbara at her home in Park Drive, and thus began one of the strongest friendships of my life. Her inner strength and spirituality helped me and so many others – Denis Corboy, who had at one stage planned to be on the flight, suffered terrible guilt and was initially inconsolable – to cope with painful grief. I found myself going back to Park Drive almost every day that summer, really getting to know Barbara and her children. By September we had asked Barbara to be godmother to the child I was due to have in early October.

My mother was preoccupied with ensuring that we would have the biggest pram and the best of everything for this grandchild of hers. She brought me in to see the formidable matron of Holles Street – the national maternity hospital in Dublin.

While Mummy was busy impressing upon her how important I was, 'a senator *and* a professor' (as I cringed quietly beside her), Matron seemed much more interested in the whereabouts of my brother Aubrey, who had recently done an internship in the hospital. We were all used to women asking after Aubrey – he was undoubtedly handsome, a charming rogue – but there was a definite chill in this matron's inquiries. When I later mentioned it to Aubrey, he gave me a sheepish grin. 'She might think,' he admitted, 'that I bumped into her parked car. I've been avoiding her!' This, I thought, was all I needed: a hostile matron.

I dutifully attended my gynaecologist and did all the prenatal classes. Nick and I had some friends around for supper at our home on the evening of 1 October, two days before my due date. To my surprise I began to feel the contractions, and it became apparent that I would have to stop cooking (with Nick's help my repertoire had moved beyond '*huevos* Maria'). I downed tools, left our guests to it, and Nick drove me to Holles Street. When we got there, we both naturally assumed that Nick would be allowed to stay with me, but the duty sister was adamant that he could not. 'You'll be the first to know,' she said, 'but no fathers are allowed in the delivery room.' We were both disappointed and upset. But then, as Nick was saying his goodbyes to me, one of the younger nurses quietly told us that the night sister, who would be coming on duty soon, would be much more flexible and that Nick should slip back in then. So Nick was present at the birth of our daughter, and I was desperately grateful for the support. One of our first visitors was my own mother, pleased as Punch when we told her that we were naming our baby Tessa.

I couldn't take my eyes off my newborn baby, and enjoyed the intimacy of breastfeeding after a tentative start. I found out – when one of the nurses asked me to express some of my milk so that it could be analysed – that I was the only mother in the entire hospital for those few days who was breastfeeding her child. I was astonished. But, I was told, mothers were embarrassed, particularly in the public wards, to expose their breasts to feed their children, and even the nurses were more comfortable when the babies were bottle-fed. Such was the culture at the time.

Nothing was ever the same again. Having a baby – with others to follow – changes everything. We had to change our routine quite dramatically to accommodate this new addition to the family, no longer able to do things spontaneously such as meet after work to go to a restaurant or socialise with friends. Nick was a typical adoring father, taking many photographs but less keen to deal with dirty nappies. The broken nights and sleep deprivation were difficult to adjust to, of course, but Tessa was a relatively easy baby, and we were lucky to have secured the services of a young nurse from Holles Street, who moved in and provided home help, allowing me to go back to work relatively quickly.

I was soon to learn of the reality for less-fortunate mothers when I was approached around that time by Maura O'Dea, an activist single mother from Dublin who had spoken out on a radio programme (and later figured on the *Late Late Show*) about being a single mother in Ireland. No one was doing this at the time. This was still quite a taboo subject: single mothers were treated as outcasts in Irish society, considered sinful, often bullied into putting their children up for adoption by family and the Church. Maura had just set up an organisation of and for single mothers called Cherish, and she wanted me to act as its president. 'I'm too young to be president of anything!' I joked, but she answered fairly sharply, 'Look, Mary, we need your name and the respectability that will give us. We're determined to make this organisation work.' So I accepted. They had asked Bishop

My grandfather H. C. Bourke, a huge influence in my choosing to study law and use it as an instrument for social change.

My grandmother Eleanor Bourke née Macaulay: unusually for a woman, master of the North Mayo Harriers.

Aubrey de Vere Bourke and Tessa née O'Donnell, medical doctors and loving parents, equipped for the beach at Enniscrone.

Dressed up for my birthday, with an enormous bow in my hair.

Trigger-happy: with younger brothers Henry and Adrian and our mother Tessa. Henry was invariably my sidekick, Robin to my Batman, while poor Adrian was set upon as the alien baddie.

An important person in my life, Annie Coyne (left), seen here with Henry and me. She was our nanny and, subsequently, nanny to my children.

Being called to the Irish Bar, 1967

Two days before our wedding in December 1970, Nick and I receive our MA degrees from Dublin University (Trinity College).

Our only wedding photograph, as we set off in a Rolls-Royce from Dublin Airport Church to the airport check-in.

With my father at Pont du Gard in the summer of 1972. He became a wonderful holiday companion for almost thirty years, and the quintessential Grandpa.

One of many happy holidays in Provence with Garret and Joan Fitzgerald and on this occasion with the writer Mary Lavin (right). At each breakfast Garret kept a fastidious account of everybody's petty cash obligations.

Proud mother: enjoying my daughter Tessa, 1975.

Beside Georges Vedel who speaks at the publication of the Vedel Committee's report in 1972 on the enlargement of the European Economic Community. The gender imbalance on the committee was not then unusual.

With my father and the Reverend Professor F. X. Martin in the front row of a Papal audience in St. Peter's Square. FX was thrilled when we invited him to take Nick's place and pretend he was 'Mr Robinson'.

At my desk, around 1980. Note my terrible habit of doodling on the box files as I talked on the phone.

Dick Spring, leader of the Labour Party, invites me to contest the 1990 presidential election.

With Limerick parliamentarian Jim Kemmy at the outset of the presidential campaign. When Jim told me I needed to smarten my appearance I had to take him seriously.

Nick took this photo as John Hume, a hero of the Northern Ireland peace process, helped me canvass in North Donegal near my mother's home town of Carndonagh.

Promising the sun – like all good candidates! – to a wonderful crowd who had braved the weather for the campaign launch in my native Ballina, County Mayo.

Áras an Uachtaráin, the President's House, with the light I placed in our kitchen window for the Irish Diaspora around the world.

My inauguration as seventh President of Ireland, 3 December 1990. I signed with the quill originally used by the third President, Éamon de Valera.

Eamon Casey to be their patron, and he – rather bravely, I thought, for a priest in Ireland at the time – agreed.

I really admired the courage, independence of mind, and leadership qualities of Maura and the other single mothers who came together as Cherish; they determined to form their own organisation rather than have middle-class women run a charity for their benefit. I remained president of Cherish for almost twenty years, until I was elected president of Ireland. I brought my children to the organisation's Christmas party every year. I valued the way Cherish succeeded in pushing for a 'single mother's allowance' to be included in our social welfare code. Cherish also helped when I and Senator John Horgan and others were seeking to introduce a bill to remove the discrimination against children of unmarried mothers, then still referred to as 'illegitimate' children. We withdrew the bill once the government undertook to introduce the Family Law (Maintenance of Spouses and Children) Act in 1976, in which the marital status of the parents did not affect the requirement of maintenance to be paid in respect of children.

In the following spring, 1973, an election was called, and a new Fine Gael–Labour coalition formed a government, unseating a Fianna Fáil government that had been in power for more than ten years. For me this meant another Senate election campaign, some weeks later. I called upon my previous supporters and some new ones, set out my platform, and prepared to seek a new mandate from the Trinity electorate.

To our family's great joy, Adrian and his wife, Ruth, had a baby girl, Elizabeth, born on 9 March. I went with my mother to see her. Mummy was thrilled with Elizabeth and spoke expansively about how many more grandchildren she would have. Two days later, on the morning of 11 March, a Sunday, she telephoned me from Westland Row. She was preparing sandwiches for my campaign volunteers, she told me: she had everything in hand and I was to concentrate on the campaign work. 'I'm just going to go across the road to Mass,' she said, 'and then I'll finish up the sandwiches for lunch.' Those were the last words I heard her speak, and they may have been her last. She suffered a massive heart attack and died there in the flat. Oliver and Aubrey, both doctors, found her and desperately made every effort, tried every procedure, to revive her. They were utterly distraught when they came to tell me the news.

My father was on his own in Victoria House and had to drive up to Dublin. Whereas we were openly weeping, he arrived in a state of calm that steadied us. He was the doctor, familiar with the suddenness of death and what it meant; he'd seen it all before, death as a natural part of life and was devastated, but feeling for us, his young.

We drove down to Ballina the following day, in a long, slow funeral procession. When we reached Maynooth, about fifteen miles out of Dublin, we came upon a group of people who had lined up along a bridge that crossed the main road. They saluted the coffin as it passed beneath. This moved me profoundly, in a wave of emotion that I still feel today as I recall it. Hours later, as we entered Ballina and proceeded through the town, in street after street the lights of premises and cars were dimmed, not all at once, but as we passed, and it seemed that every citizen of Ballina was standing outside his or her home to salute the passing of my mother. Then, getting out of the car to walk behind the hearse as it passed across Ham Bridge to the cathedral, came the sound of hundreds of footsteps behind us.

My mother had suffered from a mysterious thyroid illness and had high blood pressure for several years before she died. Even in the shock and dismay at her death, something in me was comforted by the knowledge that she did not have to suffer through what could have been a long and drawn-out illness before she died; rather, she had died suddenly, in the thick of it, full of organising everyone.

I have an image of my father sitting in the smoke room at home and seeming so bewildered and lost and sad that I wondered how he would cope at all without our mother. She had run his home and his surgery so smoothly; I feared he would fall to pieces. But he adapted to his new life stoically and came to relish the impact of grandchildren beginning to arrive in our various households. He learned to relax and became infinitely more benign.

His sense of wonder came to the fore, and he relished opportunities to travel abroad. I noticed it first that summer. Nick and I had been on a working trip to Brussels, and then joined a group of our friends, including Garret and Joan FitzGerald and Denis Corboy, in France where they had taken a house in Les Arcs for a couple of weeks' holiday. Barbara Sweetman was to join us, bringing from Ireland our baby, Tessa, now ten months old. When we asked my father if he, too, would like to come, he jumped at the idea. From then on he joined us on most of our family holidays, and it was always a happy addition. He developed, among other skills, a canny knack of staying out of parental rows and keeping his own counsel on child-rearing issues. He became a terrific grandfather, ultimately forming a particular bond with each of his twenty-one grandchildren.

I was now pregnant again. A great good-news distraction, but because of my mother's death, I got the dates all mixed up and believed that my due date was at least a month earlier than it actually was. So while I was theoretically expecting the baby in early December, not until January did I eventually go into labour, during a particularly forceful storm. On 11 January 1974, William was born. We were delighted to have a boy. 'A girl and a boy,' the nurse told Nick, 'a proper gentleman's family.'

Some months earlier, in May 1973, I had topped the poll in the Trinity Senate election. Regaining my seat gave me a surge of adrenaline, like an electric force propelling me forward. My constituents were with me, my mandate was consolidated and my knowledge of how to be an effective senator was growing all the time. But I was beginning to feel the lack of a proper economic understanding. It had never been a strength. I had been more focused on reform of law, on education, on Ireland's place in Europe. Now increasingly I was seeing inequalities in Irish life and listening to others debate how we should address the underlying financial and economic issues. My own economic thinking was not very sophisticated so I needed to look to others to deepen my understanding, and this included looking at the policies of established political parties.

Neither of the two larger parties, Fianna Fáil and Fine Gael, attracted me in terms of economic approach; their policies seemed virtually interchangeable, only personalities and their historic perspectives distinguished them. I had close friends within Fine Gael, but it was clear to me that the only political party I could join, from the point of view of its having, as far as I was concerned, the fairest economic policies, was the Labour Party. Its ranks at the time included powerful intellects such as Conor Cruise O'Brien, David Thornley, Justin Keating, John Horgan, Ruairí Quinn, Frank Cluskey – people with whom I could engage and find common ground – and it put forward the platform that 'the seventies will be socialist'. I took up the language of socialism because I believed in social democracy, and thought the Labour Party policies were fairer on issues such as education and health care and that the party's economic policies would be equitable and more distributive.

I joined the Labour Party in 1976, and when an election was called in 1977, I sought the party's nomination as candidate for the Dáil in the Dublin constituency of Rathmines West. Standing for election to the Dáil seemed a natural next step as my political ambitions broadened. I wanted to make a more substantial contribution, as part of a group of like-minded people, than I could as an individual senator. But I hadn't taken into account the intricacies of political party machinations or that other prospective candidates would, perfectly legitimately, have their eye on that constituency. The complicated selection process debarred my nomination through the ordinary channels because of a timing issue, but allowed for me to be added to the list by the party's administrative council. I suspect that the imposition of me as candidate by the party hierarchy might have caused some resentment from my Labour running mate, Michael Collins, a long-standing local councillor.

There followed some bitterness among my supporters and those of Michael Collins. They felt I was a carpet bagger imposed by headquarters; we felt that Collins had undermined the possibility of Labour gaining a seat by not ensuring that I benefited sufficiently from the transfers of his second-preference votes. The result: Fianna Fáil's Ben Briscoe scraped home on the last count, a handful of votes ahead of me, to hold onto his seat.

Initially I was despondent and felt somehow rejected, having gone all out. It had been a bruising experience, and it was a shock to my ego that, despite my high public profile, I hadn't been elected. I was also exhausted and longed to take a break to be at home with Tessa and William and make up for missing time with them over several weeks' electioneering. But there wasn't time to be at home, or to wallow: if I wanted to stay in parliament I would have to turn around and fight a difficult Senate campaign – under the Irish system, the Senate election is held a few weeks after the Dáil's. Difficult, because I was now running as a Labour senator. This meant toeing the party line and being more inhibited in what I could say. Furthermore it ran counter to the conviction of many Trinity voters that their constituency representatives should be non-party. So we knuckled down and fought a different type of campaign, one in which I appealed to the electors to understand my position and give me their support. As I stated in my election literature, 'I believe I have a constructive role to play in continuing, as I have done for the past eight years, to scrutinise legislation carefully, to speak out doggedly on minority issues, to advocate where there is need for reform and to point out injustices and discriminations where they exist.'

We lost some of our initial supporters; I no longer topped the poll, but to my relief I was re-elected. I still had a platform and a public voice. I could take some precious time to relax and be with my children, and I felt encouraged by Nick, who was constantly trying to get me to ease up. 'Look on the bright side,' he said, 'at least you won't have to be out every night at some public event.'

7

Balancing Family and Work

*But there's a way of life
that is its own witness:
Put the kettle on, shut the blind.
Home is a sleeping child,
an open mind.*

Eavan Boland, 'Domestic Interior'

LIKE MOST women, in many different circumstances, I thought constantly about work–life balance: how to be a good wife and mother, and at the same time fulfil my own potential. I had become the main breadwinner, something Nick never had problems with, and he gave me the emotional support that helped me trying to achieve that balance. Of course, there is no way I managed more than a pass mark. Though a loving mother, I was hopeless domestically. Fortunately, we had Nanny, who ran our home like a tight ship. One of the things that bound us together as a family was eating breakfast and our evening meal together, and we started a tradition of having Nick's aunt Audrey, granny-figure to our children, to lunch every Saturday.

Deep inside me, I knew I was lucky that I woke up every morning looking forward to the case in court, or the debate in the Senate, or to preparing a lecture for my students. The work was fulfilling, and at times I had the satisfaction of making a real difference in people's lives, a small breakthrough for justice and equality.

At the same time, I felt intense joy at watching the way Tessa and William socialised together – played, fought, competed, and learned – as they were close enough in age to be real companions. My own relationship with my brothers remained very close, so our various children developed a strong bond as cousins. This was reinforced by their relationship with Grandpa (as my father was now firmly known in the family). I also enjoyed a very good relationship with Nick's father, the Boss, and that younger generation of Robinson cousins who were also very close to our two young.

The teaching salary I received as a part-time lecturer was rather low, but together with my salary as senator, it allowed me, as I had hoped, some freedom to choose the cases I would take on. From the mid-1970s I no longer went down on the court circuit to the West of Ireland with John Willie O'Connor; I had learned a great deal from sitting in court with him and listening to his principled and locally well-grounded approach. I now based my practice in Dublin.

There were some satisfying wins: I represented Máirín de Burca and Mary Anderson in their successful challenge of the laws that effectively prohibited women from sitting on juries, and overturned a law that prohibited a couple of mixed religion from adopting a child.

I was neither a full-time nor a mainstream barrister: I turned down some cases that I wasn't interested in – 'slip and fall' or 'running down' personal injury cases that were the lawyer's bread and butter. Because I didn't have a financial need, I could pick and choose, but that did not endear me to some of my colleagues, and I was never fully at home in the rather conservative, male environment of the Law Library. Still, I felt confident in what I was doing, and because I had an income from teaching and the Senate, my primary motivation was not a drive to make money from my legal practice.

At the benching of a senior colleague (an honour conferred on a senior barrister by members of the judiciary and other benchers at a formal occasion held in the King's Inns), I got into a heated argument with a group of male colleagues. My argument was that so many briefs of the straight-forward personal injury type (of which, of course, a large percentage settled before trial) were handed over from one barrister to another, at the last minute. The busier practitioners found themselves double or even triple-booked but held on to their cases in the hope that one of them would settle. My female colleagues were adamant that they were losing out on these cases because of *where* they were being handed over: in the men's robing room. Few of the men understood my point. The practice was such an inherent part of the Law Library culture at the time that they were blind to the fact that it discriminated against women barristers. (I was comfortable making this argument, as I was not looking for these types of briefs.)

Winning what seemed like hopeless cases on behalf of marginal-ised, stigmatised clients did not always endear me to some colleagues, either. The travelling community in Ireland suffered (and, in some senses, continues to suffer) from terrible prejudice and discrimina-tion. I had taken an interest in the 'problem' of travellers while still at university, and was always open to taking on cases on their behalf. A particularly defining case was that of Rosella McDonald. Dublin City Council had attempted to remove Rosella and her family of ten children, living in two caravans, from their unauthorised camp on the works site of a bypass road. The High Court ruled that travel-lers could not be evicted from local authority property without being offered a suitable alternative. The judge, Donal Barrington, stated, 'Nothing is solved merely by moving such families from place to place. By doing so, not only is the problem perpetuated but the claims and rights of the children to any possibility of education and a settled life and future are ignored.'

This statement was a wedge in the door, acknowledging as it did a duty of care on the part of local authorities to provide for the housing needs of travellers that trumped the local authority's right as property owner to eject 'trespassers' from its land.

I also took a case on behalf of a working married couple whose claim that they were being discriminated against by being taxed more as a couple than they would be as two single working people was upheld by the High Court. When, as a senator, I discovered that the rules governing pensions for spouses were different for male and female senators (when a man died, his widow and children became entitled to his pension benefits, but when a woman died there was no comparable pension or allowance for her widower and chil-dren), I took action to fight this discrimination. I asked Pat Rabbitte (now, as I write, minister for communications, energy and natural resources), as my trade union representative, to argue the matter before an equality officer under the Anti-Discrimination (Pay) Act 1974. It would have appeared, nowadays, to be a 'slam-dunk' case, but in fact we had to pursue it as far as the Labour Court, the equality officer having found that the differentiation between male and female senators occurred after death, and therefore did not arise during the course of employment and so could not be investigated under the legislation! The Labour Court overturned this interpretation and held that I was entitled to the same provision in relation to benefits for my dependants as applied to male senators, a ruling that would benefit all future women senators. It gave me great satisfaction to achieve greater equality in my own work-place.

But it was a planning case, one concerning Dublin City's heritage and origins, that would occupy much of my professional time, and Nick's, during this period.

Now practising as a solicitor, Nick had gained a reputation as a conservationist interested in preserving Irish heritage, having, among other things, taken on the King's Inns (and thereby the benchers of the King's Inns: the judiciary and senior members of the Bar) in their controversial decision to sell off a valuable collection of non-legal antiquarian books held in the King's Inns Library. In 1976, he and our close friend Edward McParland had established the Irish Architectural Archive, a national record of Ireland's built heritage, from the earliest structures to contemporary buildings, and had found premises in which to house the archive and make it available for public use, in the heart of Georgian Dublin.* So when a calamitous threat to the historic fabric of the city arose, Nick was an obvious person to call upon. He received instructions to act from FX Martin, a priest of the Augustinian order and a professor of medieval history at University College Dublin, who had formed a group called the Friends of Medieval Dublin in response to the proposed building of civic offices by Dublin Corporation on a site at Wood Quay, between Christchurch Cathedral and the River

---

* Now, supported by the state, its magnificent collection is housed in a double-fronted house on Merrion Square.

Liffey. The site had been compulsorily purchased by Dublin Corporation for this purpose, but when excavation started, not only were valuable archaeological remains from the Viking period discovered but also the earliest settlement of what would become the city of Dublin. Nick briefed me as junior counsel. As FX told us at an initial consultation, the remains included the ruins of part of the old city wall, Viking-period defences, banks, and palisades, the first of their kind to be discovered in Ireland and unique in western Europe. Unless works on the site ceased, the foundation of the city, its very footprint, was in danger of imminent destruction.

Accompanying FX at that consultation was a young school teacher and active member of the Friends of Medieval Dublin, Bride Rosney. When we set about issuing proceedings seeking an injunction restraining the works and a declaration that the site was a national monument, Nick and I could see that while FX was compelling on the reasons why we needed to save Wood Quay, the real organising presence was Bride. Highly intelligent, with a terrific sense of humour, she was incredibly reliable in following up on anything we ever asked of her. As junior counsel I was responsible for conducting research, drafting pleadings, and gathering together the evidence and the materials to exhibit in the affidavits we would be relying on in the case, including those of a number of eminent, internationally recognised historians and archaeologists. Bride's input was invaluable. Nick persuaded Donal Barrington, who had led me in some of my earlier constitutional cases and had not yet become a High Court judge, to act as senior counsel.

Preserving the medieval heart of Dublin was something both Nick and I felt strongly about; it would prove difficult at times to maintain the dispassionate demeanour of a legal professional arguing on behalf of a client.

The issue had garnered a great deal of interest among Irish and international experts and scholars, who were salivating at what they described as one of the most important archaeological sites of its kind, and also among ordinary Irish citizens, particularly Dubliners, who rightly viewed this as a unique part of their heritage and culture.

Donal Barrington argued the case astutely, and in June 1978 the President of the High Court, Mr Justice Liam Hamilton, declared the site a national monument. FX Martin, the Friends of Medieval Dublin, many supporters of the campaign and the ordinary citizens of Dublin celebrated this victory and defence of their cultural heritage.

But the celebrations were short-lived. Less than two months later, and to public shock and dismay, Dublin Corporation invoked a loophole in the National Monuments Act 1930 that permitted them to destroy a national monument with the formal consent of the Commissioners of Public Works. Having been frustrated in so many previous attempts to acquire and develop a site, Dublin Corporation was desperate to build a new, centralised civic office, and wanted to pursue this site rather than start the whole process again elsewhere. A period of six weeks was granted for archaeological excavations to finish before building would commence.

The Friends of Medieval Dublin organised a protest march. In FX they had a charismatic leader, absolutely determined that Wood Quay would be preserved as part of the heritage of Dublin and Ireland. In a huge turnout, some twenty thousand marched. The atmosphere was electric and for a moment we collectively believed that we could make a difference, and through this show of public opinion sway Dublin Corporation to alter its decision. It seemed incredible that we had to fight our own short-sighted local authority to preserve this national treasure. Addressing the crowd that Saturday afternoon I said, 'Wood Quay has taught us a very valuable lesson. We have learned that democracy demands more of citizens than a passive role in between elections. It is not enough to vote every few years and then sit back and let all decisions be taken by a small group. There must be constant vigilance. Where necessary there must be active participation and involvement in order to protest. That is what we are witnessing today. We are witnessing a huge protest by ordinary people in Dublin against a decision taken in their name to destroy a national monument. This peaceful protest is part of the lifeblood of democracy, and its very size shows that the pulse is strong . . .'

The focus of the campaign, then, was to persuade Dublin Corporation to seek an alternative site for its offices. FX brought another legal action, this time challenging the validity of the joint consent, the loophole permitting the destruction of the site. We were thrown out by the Supreme Court. The corporation did agree, periodically, to extend the deadline for completion of the archaeological excavation, but time was running out and tensions were running high between the construction workers and the archaeologists.

Local elections were imminent and it was hoped that a change of personnel would change the corporation's mind. I agreed to run for Dublin City Council as a Labour Party candidate on this single issue, and thanks mostly to Wood Quay activists canvassing extensively on my behalf, was comfortably elected.* I was delighted to do it. I didn't have a lot of spare time, but I felt so strongly about the case, which had become a personal mission for Nick and me, and I wanted to do anything in my power that might make a difference.

In the meantime, in a dramatic move intended to 'hold' the site until after the election, students and scholars, citizens eminent and ordinary – including historian Kevin B. Nowlan, writer Mary Lavin, sculptor Oisin Kelly, publisher Michael O'Brien, and our clients, FX, Bride and others – took physical occupation of Wood Quay.

I visited the site regularly during the occupation, but in my heart I knew the occupation had no legal justification. After two weeks the construction company, John Paul and Co., inevitably brought proceedings in the High Court seeking an injunction to remove the occupiers. To the astonishment of all of us, really, Mr Justice Gannon found reason to refuse the injunction, siding with the occupiers, and his ruling was immediately appealed to the Supreme Court.

I remember well the scene in the Supreme Court that morning. The judges, led by Tom O'Higgins, the Chief Justice, came out onto the bench, brandishing copies of the Prohibition of Forcible Entry and Occupation Act. They were furious with Seán Gannon for refusing the injunction in the High Court. They were furious that the case was still dragging on. And they were furious with me; I

---

* Before the abolition of such dual mandates in 2003, it was not at all unusual for a senator or indeed Dáil deputy to seek election to a local authority.

think they saw me not so much as legal counsel but as a pushy young public representative. They certainly closed down very quickly any argument I tried to present, and then directed me personally (a judicial rap on the knuckles) to go immediately to the occupied site, directly across the River Liffey from the Four Courts Building, and tell my clients to leave forthwith; that that was the order of the court and if they refused they would be liable to criminal prosecution. I was then to come straight back and report to the court. Which, red-faced with humiliation, I did.

It was a sorry moment and a heavy responsibility, advising the occupiers that they had to leave, that however angry they were, it was not worth a criminal conviction.

As a newly elected member of Dublin Corporation I did my best, along with others, to keep the Wood Quay issue alive, writing letters and bringing motions and so forth, but the most we could achieve was to extend the time for the archaeological excavations to take place before demolition and construction. It was like banging one's head against a wall, trying to explain to the senior civil servants just how much damage they were doing, and the significance of this lost opportunity. We argued the example of the English city of York, where it had been demonstrated that buildings could be erected with modified foundations that facilitated long-term excavation and display of the city's origins, attracting visitors from all over Europe. None of our entreaties as city councillors made any difference. They were just so desperate to build their new offices; they had no concept, whatsoever, of the value of the site, seeing it purely as a development opportunity and the archaeology as the recovery of items among the lorry loads of earth and rubble that could be sifted through at another location. As things grew more heated, councillors were threatened that we would be personally surcharged with the very considerable costs of any further delay, and thereby bankrupted.

Huge regret was felt by the ordinary people of Dublin. Young Irish archaeologists were in despair; they knew they could have spent their whole career on so rich and unique a site.

The diggers took over, construction began and the civic offices were duly built, great big concrete blocks of buildings. For a long time, years, when driving or walking in the city I would try to avoid looking at the buildings. I just didn't want to relive the sorrow.

During this time, Nick and I became very close to Bride Rosney and FX Martin, forging friendships that went way beyond the Wood Quay case. Once, when we were invited by mutual friends to a drinks party, and their printed invitation indicated a dress code of 'formal or clerical' some of us conspired, as a practical joke, to choose the 'clerical' option. Borrowing robes from FX, Nick dressed as an Augustinian monk, Ann Lane went as a nun with full wimple and veil, Bride donned the vestments of a bishop (and swears that when she filled her car with petrol on the way to the party, the guy at the petrol station said, 'Thank you, Father' when giving her change), and I dressed up as a Hare Krishna, complete with traditional robe and bells. Our arrival coincided with that of the Protestant Archbishop of Dublin, Donald Caird, who greeted us serenely and added authenticity as we made our eccentric entrance. One by one our hosts and their guests recognised us with a kind of shocked double-take that ended in a laugh or a cheer.

In 1980 FX persuaded us and my father, with whom he had developed a close friendship, to travel with him to Rome on an occasion when he would be there on official business relating to a canonisation. FX was a spiritual man but he had a worldly aspect, too; he was tall, handsome and charming, and had a certain amount of personal vanity, wearing a black, scarlet-lined cloak with great flair. In Rome he took us to a number of inexpensive but excellent restaurants, off the tourist trail, where he would greet the chef with a bear hug and we would be treated like kings. FX would talk about his subject, medieval history, in a spellbinding way as we sampled the best of Roman cuisine. One evening, as we emerged from such a restaurant, well fed and watered, a Vespa sped past us whose passenger snatched my handbag. The stream of invective from FX to the miscreants shocked my poor father to the core.

FX had arranged for us to attend the weekly papal audience, which would fill the vast St Peter's Square. Goodness knows how he managed it, but my father, Nick and I had been allocated tickets in the first row, numbered one, two and three. We realised, of course, that FX himself would love to take part, and Nick insisted that FX take his ticket and be 'Mr Robinson'. When FX arrived that morning, out-of-breath, to meet me and my father, he was dressed in a black suit but had been unable to find a tie. Taking our places in the front row, we teased him mercilessly about this sartorial faux pas. When His Holiness John Paul II came towards us, FX, forgetting his role as Mr Robinson, insisted on introducing me as an important Irish senator, to which the pontiff replied warmly, in his heavily-accented English, 'God bless you, and God bless your function.' My father was thrilled. Looking around I caught the eye, and the thin smile, of former Taoiseach Liam Cosgrave, two or three rows behind.

The three years I spent as a city councillor were good learning years for me, although I would not count them as particularly successful. I came to understand how one ran a city; in particular, that the city manager, the non-elected civil servant, rather than the elected repre-sentatives, had the real power. As a Labour Party member I followed the party whip and this did not always sit well with me, when, for example, I found myself voting in support of lord mayors against my own better judgment. The city council had a range of responsibili-ties but little political power; the tendency was to form committees, and more committees, and these always seemed to meet at six or seven or eight in the evening when I wanted to be at home with my young family. So eventually I resigned, with mixed feelings of regret – to be leaving a position where I was learning so much about the diversity of Dublin and its people – and relief that I would have more time at home.

In late 1980 I was approached by the Labour Party to run again for a Dáil seat, this time in the Dublin West constituency. The timing was not good: I was pregnant again – something of a surprise, and a gap of seven years since William was born, but we were thrilled nonetheless. My due date was early May 1981, and the general elec-tion looked to be called around that time. Notwithstanding the considerable time commitments that would be involved were I to be elected, I found it hard to turn down a Dáil nomination, with the possibility (albeit slim in this instance as the odds were against a

Labour candidate being elected in this constituency) it presented of a higher platform from which to advocate politically for social change. So I agreed to the nomination, and then sure enough the election was called on 20 May, less than three weeks after Aubrey was born, on 3 May. Canvassing with a tiny baby was going to be tough and require a lot of organising. It was a tense election. The dirty protest was going on in Northern Ireland: IRA prisoners were no longer being classified as political prisoners and protested this loss of status by refusing to wear clothes and soiling their cells with their own excrement, in a precursor to the hunger strikes in 1981 in which ten IRA prisoners, led by Bobby Sands, would die. There were some dirty-protest candidates running for election in the South, including one in the Dublin West constituency.

I had been spending more and more time with Bride Rosney, and enjoying her company. I was struck by the strong friendship she had with her mother; they were two similar characters who occasionally had strong differences of view. Bride had the great skill of being able to get on the wavelength of children of any age. She became a close friend of Tessa and William and would regularly take them on trips to Dublin Zoo or to the cinema. When Nick and I were deciding who would be Aubrey's godmother, she was a natural choice.

Bride was not a member of the Labour Party but advised me personally as a candidate. At first I sought her opinions specifically on education issues, in which she had carved out deep expertise, but then more and more I would turn to her for advice on other issues, as I valued her political instincts and gut reactions. I couldn't have managed that election campaign when Aubrey was a newborn without her. She had a teaching job herself, but whenever she could, she would mind Aubrey in the home of one of my constituents while I was canvassing, and I would meet them there so that I could feed him. Sometimes constituents recognised me from a photo with Aubrey taken a few days after his birth, and would say, 'You should be at home minding the baby.'

It was the kind of constituency where, if you secured a vote from one person, you might get their whole extended family, even twenty or thirty votes. Bride had taught in Ballyfermot and knew the area well. She brought a young woman from one of these big families, a former pupil, in to see me. Jackie and her husband and two children lived in the parlour of her mother's house, which did not have a bathroom. There were twenty-two people living in that three-bedroom house. I sat Jackie down and asked her questions, taking notes as we talked. At the end I advised her that she certainly had enough points under the calculation used by the local authority to secure a house. I gave her the address of the housing office of Dublin Corporation and sent her up there, telling her to keep at them until they agreed that she was entitled to a house, assuring her she'd have it within a month. Bride was pleased with me: having done the business with Jackie I should get those votes.

I polled badly in the election. I think even the dirty protest candidate outpolled me. Bride bumped into Jackie afterwards. 'We all voted for Jim Mitchell [the Fine Gael candidate]' said Jackie, 'he got us the house.' That taught me something. I had tried to empower Jackie; I had gone through her situation with her, had advised her, and had told her how to get the house for herself. An experienced, well-organised politician like Jim Mitchell would have taken her name, promised to look after it for her and given it to one of his helpers to organise. He worked the system expertly and he won the seat easily.

I didn't run for the Dáil again. I wasn't exactly disillusioned; rather, I understood better my limitations in terms of playing the political game, how much, if at all, I was willing to compromise, where my priorities lay. More and more I was realising that it was through taking court cases rather than as a frontline politician that I would best be able to advance the causes I believed in.

Of course, I then had to bounce back and retain my seat in the Senate, which, happily, I did, in that election and – it was a turbulent time, politically – in the two subsequent elections held the following year.

Prior to that election, I had taken on another absorbing case, one I felt could potentially be groundbreaking. Having endured years of physical and mental cruelty at the hands of her husband, Mrs Johanna 'Josie' Airey of Cork had initiated what she believed to be separation proceedings in the local District Court in or around 1972. She had been attempting to get her husband to agree to a separation for some eight years prior to that but had failed to get the necessary cooperation from him, and of course, divorce was prohibited in Ireland by the 1937 Constitution.

Mrs Airey was bitterly disappointed by the District Court proceedings. She was looking to the courts – and to the state – to protect her and to liberate her. The District Court had convicted her husband of assaulting her – and fined him twenty-five pence – but that court did not have jurisdiction to grant her a legal separation from her husband: such power was only vested in the High Court.

Mrs Airey had been represented by a solicitor in the District Court. For the High Court she sought legal advice on the question of separating from her husband, trying five different solicitors' firms, but none would take her case. She could not pay their fee, and even if she won (and she was confident of success), with the possibility that her legal costs would have to be paid by her husband, he, although employed, did not have the kind of money in question.

Mrs Airey then contacted a free legal aid centre operated mainly by students in Cork, but they lacked the resources to represent her. She wrote to government ministers and other Irish politicians pointing out the injustice of her situation and her inability to obtain a remedy. She commenced the process to obtain a church nullity, although such nullity would not have legal effect on the status of her marriage.

Josie Airey wanted a judicial separation to protect her and her children from an alcoholic and violent husband, and to give her the security and status of a legally separated woman who could begin a new life free from apprehension and fear. Despite the fact that her husband had left the family home, she still wanted the judicial separation; she needed the permanent protection and security that such a remedy provided, including relieving her of the duty to cohabit in the event of her husband's return – which she anticipated at the time – and because she wanted to secure her succession rights and the rights of her children.

Her situation was similar to that of thousands of women in Ireland, who had found themselves trapped in an irretrievably broken marriage, with no money to obtain the legal advice and representation required to regularise (and in some instances, escape) their circumstances. What was different about Josie Airey was that she had a strong sense of her right to justice, and the courage and determination to pursue that right in every way she could. Unlike so many other dependent wives in Ireland in similar circumstances, she was not prepared to resign herself passively to the unhappy position in which she found herself.

In June 1973 she wrote a lengthy, detailed handwritten letter to the European Commission of Human Rights, a body she had read about in a local newspaper, complaining that her human rights had been violated. Tied up with her anguish over her inability to hire a solicitor to represent her in High Court proceedings, she made a further pleth-ora of claims about her husband beating her and his alcoholism, how he required treatment, and how she could not obtain a passport to travel abroad without his permission. She included a claim of brutality by the police and unfair procedures in relation to another matter that was before the District Court, where she alleged she had been hauled out of her sick bed under warrant for failing to appear to answer claims in relation to her son's poor school attendance. She had spent four nights in prison in default of paying a fine of around £3.

The remarkable thing is that this working-class woman – intelligent but not educated – somehow, without any legal advice, initiated going to the European Commission of Human Rights in Strasbourg. The beauty was that this was the perfect forum to seek redress in her case. Mrs Airey sent her letter to the Commission, and its lawyers were prepared to sift through all the irrelevancy and find that there was something worth communicating to the Irish Government: that this woman needed a separation from her husband, and that no solicitor locally would take her case.

The Commission wrote to Mrs Airey saying that it would give her legal aid to put together submissions arguing that her case was admissible before the Court of Human Rights.

The president of the Law Society, Patrick Moore, had been involved in the only case at that time that had gone to the Court of Human

Rights in Strasbourg from Ireland: a case concerning internment in the 1950s, *Lawless v Ireland*, which had been unsuccessful. The Commission, through the Irish Commissioner at the time, Brendan Kiernan, contacted Pat Moore about Mrs Airey's case, and he sent the papers to a young solicitor whom he admired, Brendan Walsh.

Brendan Walsh had been a student at UCD while I was at Trinity. We had met at various social functions and he knew my older brothers. Apparently Mrs Airey knew of me by reputation, and asked Brendan to brief me in her case. He duly called me and I readily agreed to take it on.

To me, initially, it seemed like a speculative case, difficult to win, but it was intriguing because of the paradox of Mrs Airey's being awarded legal aid by the Commission in Strasbourg to see whether she should be entitled to legal aid to bring her case in Ireland. Clearly the Commission felt there was something valid in her case. So it attracted me not only because it was an opportunity to argue human rights in the European context, but also because we realised that if we won, it was not just Josie Airey who would get her justice. A win would have implications for civil legal aid in Ireland, where there was no state-run scheme in place, and indeed in wider Europe.

Mrs Airey's claim was that the Irish state had failed to protect her against the physical and mental cruelty of her allegedly violent and alcoholic husband by not enabling her to legally separate from him by virtue of the prohibitive cost of High Court proceedings, which did not attract legal aid. I was able to frame an argument for her that this complaint was also based on Ireland's obligation under the European Convention on Human Rights (to which it had been a contracting party since 1950) to provide Mrs Airey with access to the appropriate court without any discrimination or differential treatment in order to protect her rights under family law, or alternatively, provide some other redress.

Following our written reply in response to the Government's observations challenging admissibility, the Commission decided we should present argument in person and extended the legal aid to allow Mrs Airey and her legal team to travel to Strasbourg. She brought with her Maureen Black who was president of the Cork Citizens Advice Bureau and had championed her cause. This trip to Strasbourg was itself part of the empowering process for Josie Airey. It is not always necessary for an applicant to be present for what comes down to quite technical legal argument, but as her very case centred on *access* to the courts for *access* to justice, the Commission agreed that it was important for Mrs Airey to attend in person. And so Josie maintained ownership of the case she had initiated, and attended with great pride at the hearing before the Commission. Listening to the proceedings was important for her, her sense of justice was very strong. She was quoted in the *Sunday World* newspaper as saying, 'It's really the whole principle of the thing that made me decide to go ahead and contest the matter. Women really are second class citizens and I want to prevent other women and children from going through what I did.'

The oral hearing was set for July 1977, and the four of us, Josie, Maureen, Brendan and I, flew to Strasbourg.

Counsel for the Irish Government immediately went on the attack, claiming that Mrs Airey's application was totally without merit, hypothetical and unreal, and alleging *mala fides* ('a disturbing lack of candour'). Tactically this line of attack was a miscalculation: it signalled to the Commissioners that the government had not understood that the Commission had itself instigated the formal application by ensuring Mrs Airey was professionally represented, having identified sufficient merit to do so in the story she had told them.

The Irish Government argued first, that there were summary legal remedies available to Mrs Airey in the District Court that would satisfy her claim, such as obtaining a barring order to protect her from her husband and, as she had already done, claiming maintenance orders against her husband which were 'easily' enforceable. Second, that her right of access to the High Court was in no way denied by the state and the mere fact that she had insufficient means to pay normal legal costs was not the responsibility of the government, particularly in view of the fact that it was not expressly required by the wording of the Convention on Human Rights to provide free legal aid in civil matters. She could always appear for herself without a lawyer.

The government, therefore, further argued that Mrs Airey had not pursued the remedies at her disposal – either by seeking the alternatives it proposed in the lower courts or by representing herself in the High Court on the claim for judicial separation – and that, therefore, her application was inadmissible.

The Commissioners questioned us on this point after the hearing. They wanted to establish just how many lay petitioners there had been in the last decade seeking judicial separation in the High Court. When we obtained official statistics on the matter from the Central Office they revealed that there had not been a single unrepresented petitioner for a decree of judicial separation. This put in doubt the government's argument that Mrs Airey would be capable of representing herself without a lawyer, and affirmed our point that because of her financial situation Mrs Airey's access to the court was being impeded.

In a decision of 7 July 1977, the Commission declared Mrs Airey's application admissible, finding that her complaint raised substantial issues of law and fact under Articles 6, 8, 13 and 14 of the Convention, whose determination required an examination on the merits. This was a great victory for Josie Airey. Subsequently, on that examination, the Commission was of opinion that the failure of the Irish state to ensure Mrs Airey's effective access to court to enable her to obtain a judicial separation amounted to a breach of Article 6, and it took her case against Ireland to the court in Strasbourg (which she again attended with us). The Commission's principal argument was that while in theory Mrs Airey could appear in person in the High Court to plead her own case, it was unreasonable to expect a person untrained in the law and procedure associated with judicial separations in Ireland, and so intimately affected by the issues involved, to act on her own behalf.

There was a flurry of media interest, with Josie described in the *Irish Independent* as, 'The woman from Cork, who has made legal history by becoming the first woman to bring the Irish Government this far in a case . . .' Commentators in all the broadsheets predicted that the newly elected Fianna Fáil Government would submit to the international pressure then mounting to bring in free legal aid in family law cases in Ireland and introduce, as promised, informal and less institutionalised procedures and tribunals to deal with family law cases, in order to avoid a court ruling from Strasbourg.

The government decided to fight the case, and, when the court ruled in October 1979 in favour of Mrs Airey, was required both to set up a system of civil legal aid and to simplify family law procedures.

The editorial in the *Irish Times* the following day was a resounding vindication.

> Mrs Airey's success at Strasbourg is much more than a one-woman triumph over the system. Yesterday's judgment will alter, in a quite fundamental way, the relationship between our civil law and the ordinary citizen of limited means. It has social implications which, in time, will change the lives of a great many people . . .
>
> If there is a greater evil than the ending of a marriage, it must be the continued imprisonment of a wife or her children in a home where violence and degradation are the norm. Strasbourg – and a determined Mrs Airey – have ensured that it need happen no longer.

Mrs Airey died in 2002 at age 70. Her legacy was not simply the considerable achievement of obtaining in the European forum redress for the injustices visited upon her, but the opening up of legal systems within the wide Council of Europe area to ensure that access to domestic courts could not be frustrated by a claimant's lack of means.

This was the type of case I loved, where an individual client had the courage to seek redress for injustices caused by acts or omissions of the state, and where the result impacted the entire system, or positively affected many other people in similar situations.

A few days after the court's judgment I received a short, handwritten note from Tommy Conolly SC. 'Congratulations on your most notable achievement in Europe in the legal aid and family law fields. Future generations will regard this as a milestone!' I was most encouraged by those words, Tommy had been a hero figure for me when I was starting out on the Western Circuit and there was no one in the Law Library whose esteem I valued more.

8

Windows of Change

It is difficult
to get the news from poems
yet men die miserably every day
for lack
of what is found there.

William Carlos Williams, 'Asphodel, That Greeny Flower'

IN AUGUST 1982 Nick and I packed our three children, Nanny and Grandpa, and we set off to California for the month, where I was going to teach European law to summer-school students at the University of San Francisco. It was not a heavy workload, two teaching sessions of two hours each week in the evening. Rather, it was a chance to have a really good family holiday. We rented a house in a neighbourhood called Walnut Creek across the Bay Bridge from San Francisco, enjoyed the lifestyle, and caught up with friends of ours living in the area.

On one occasion Nick had flown to the East Coast to have a number of fund-raising meetings for an Irish conservation body, the Heritage Trust, and I flew over to join him for an event that he was speaking at, leaving Grandpa and Nanny in charge. My father's driving even at home in Ballina was pretty iffy (his spectacles were as thick as jam jar bottoms by this stage), so I warned him to stick to the lesser roads and not to go out on the highways. One afternoon he was driving the children back from the nearby park where they had all been swimming and he found himself in the wrong lane to turn left, but went ahead and turned anyway, crossing two inner lanes of traffic as he went (he never would have found his way home otherwise). Immediately a siren sounded, and a police officer on a motorcycle pulled him over. The officer approached the car, stern-looking, taking off his gloves, pulling his notebook out of his pocket, and Grandpa rolled down the window, with a slightly puzzled, concerned smile on his face, as if to say, Is something the matter? 'Sir, I'm going to have to give you a citation,' said the officer, to which Grandpa replied, 'Gosh, that's very kind of you. What am I getting the citation for?' Hearing his accent, the cop looked at him with a grin, 'You're Irish!' he said, and let him off with a warning and a funny story to tell us on our return.

Nanny had been complaining about a sore tummy while we were in California. It was unlike her to complain at all about anything to do with herself, and as soon as we returned, I took her to the doctor, who examined her swollen stomach and referred her straight to hospital. I was called in by a senior nurse who told me they'd found a tumour and it was bad. Nanny, who was in her late fifties, went through chemotherapy, it failed to kill the cancer, and weakened her terribly. She went home to her family in Mayo. She clearly was unhappy there; she felt like a burden on her mother and sister, she was lonely, and she missed the children (and they her). There was no question, of course, of her ever working again, but when I asked her whether she would come home with me, and persuaded her that we would work it all out together, she agreed. It was, after all, her home, more than anywhere else. We had by now hired a young woman, Laura Donegan, to help out. Nanny would sit on her chair in the kitchen, still a commanding presence, and, in a sense, run the house from there, telling the children – and, no doubt, Laura – what to do.

As she got frailer I bathed and dressed her. Our relationship had gone full circle; this woman had washed me, nurtured me, and then helped raise my own children; now I could try on a tiny scale to give back to her. Nanny stayed with us until a matter of days before her death, when she simply had to be moved into palliative care in Ballina, where her own family could be with her as she peacefully passed away.

Nanny's absence left a gaping hole in our home as we mourned her and remembered with affection her funnier moments, such as her urging Nick to have another helping of plum pudding, 'Sure, it's only fruit.' But our smiles were sad, and we found it hard to comfort our young children on this their first loss of a loved one, and to explain to them the permanence of death.

I had taken silk in 1980. This means I moved from being 'junior counsel' to 'senior counsel', the 'silk' referring to the fabric of the distinguishing legal gown I would now don in court. Junior counsel's role is to research, draft pleadings, appear in pre-trial motions and do some arguing in court. I felt ready now to step up and take the lead in arguing cases in court. I was one of only a handful of women silks, but I felt I got a good run from the judges and was treated equally and, generally, with respect by my male colleagues.

A number of commentators predicted that Mrs Airey's case would have implications for the constitutional ban on divorce in Ireland at the time. But the question of divorce was never an issue in that case; rather, it was about accessibility to the remedy of a legally recognised dissolution of marriage without the right to remarry. In a slightly later case, *Johnston v Ireland*, I had instructions to pursue the right-to-remarry argument.

Roy Johnston, his partner, Janice Williams-Johnston, and their daughter lived together as a family, but the Irish law would not recognise them as one because Roy had not ceased to be married to someone else. That marriage had irretrievably broken down, and the parties had obtained a judicial separation many years before Roy and his new family took this action, but the Irish Constitution at the time prohibited divorce, Article 41.3.2 providing that 'No law shall be enacted providing for the grant of a dissolution of marriage'.

Not only did this mean that Roy and Janice were prohibited from marrying and from enjoying the usual rights that come with marriage such as succession rights and family home protection, but it also meant that their daughter was considered 'illegitimate'. This had real effects: Roy was prohibited from acting as guardian of his daughter; she had no right to inheritance were he not to make a will, nor would she have a claim against his estate in the event that he failed in his moral duty to make proper provision for her under a will.

In Irish law, the 'family' that was afforded protection by the Constitution and laws was only and always the family based on marriage.

Roy and Janice instructed me, through their solicitor, Mona O'Leary, to seek redress under the European Convention on Human Rights, and we lodged an application against Ireland in 1982, including their daughter as a third applicant.

The couple claimed that the absence of provision for divorce in Ireland violated their rights under the Convention, primarily Articles 8 and 12. Article 8.1 provides that 'Everyone has the right to respect for his private and family life, his home and his correspondence', and Article 12 provides that 'Men and women of marriageable age have the right to marry and to found a family, according to the national laws governing the exercise of this right'. Out of these two articles we sought to extrapolate a right to remarry, and therefore a right to divorce.

It was a leap, and one the European Commission on Human Rights did not take, deciding in May 1985 that a right to divorce and subsequently remarry was not guaranteed by the Convention. The Commission did, however, find a breach of Article 8 in relation to the status of the third applicant, the Johnstons' daughter, holding that because of the legal regime concerning her status, Irish law had failed to respect the family life of the applicants.

The Commission referred the case to the European Court of Human Rights, and an oral hearing was held in Strasbourg in June 1986. That hearing, before twenty-one judges of the Court, attracted media attention that in turn brought home to me the courage of this family. In Irish family law a strict in camera rule is observed to ensure that parties to family law proceedings are not identified. Here, in the European Court of Human Rights, Roy and Janice were utterly exposed. They sacrificed their privacy and that of their daughter to pursue reform of what they considered, through their personal experience, unjust law. They could do this partly because, years before, Roy and his then-wife had so maturely dealt with the breakdown of their marriage. Still, the exposure of their private life must have been difficult to bear at times.

Roy and Janice became figureheads of the pro-divorce campaign. Their case was supported by the Divorce Action Group. Many couples in Ireland were in the same situation, where one partner could not find release from an earlier, irretrievably broken-down marriage; was now in a long-term stable relationship, with children; and could not regularise that relationship or have it recognised as a family in the eyes of the law.

Once again Irish law, influenced by Roman Catholic doctrine, was not reflecting the reality of people's lives. As with the huge instance of 'menstrual cycle irregularities' in the Irish female population in the context of the ban on contraceptives, the Irish law on nullity of marriage (whereby a court declares the marriage invalid, null and void from the outset, as distinct from divorce, where the marriage would be dissolved) was the most availed of in the Western world. It was the only way to get out of a marriage that was no longer sustainable.

The divorce referendum in Ireland, held in the same month as the oral hearing in *Johnston v. Ireland* in Strasbourg June 1986, was emphatically defeated: 63 per cent to 37 per cent against a change in the law. This was in no small part due, I believed, to dishonest campaigning and fear-mongering by the anti-divorce groups. Less than a decade later, in November 1995, after another hard-fought campaign, the Irish people voted by the narrowest of margins to remove the prohibition on divorce. Divorce legislation has been on the Irish books now for some fifteen years, and the sky hasn't fallen in.

In Strasbourg in December 1986, the European Court of Human Rights delivered its ruling in the *Johnston* case: a right to divorce (or a right to remarry) could not be derived from the right to marry contained in Article 12. However, the Court went on to hold, first, that the applicants clearly constituted a 'family' for the purposes of Article 8 and were thus entitled to its protection, and second, that Ireland had breached their Article 8 rights in the way it discriminated between the status of the child of their family and the child of a marital family.

The Court said, 'In the present case the normal development of the natural family ties between the first and second applicants and their daughter requires, in the Court's opinion, that she should be placed, legally and socially, in a position akin to that of a legitimate child. Examination of the third applicant's present legal situation, seen as a whole, reveals, however, that it differs considerably from that of a legitimate child . . . the absence of an appropriate legal regime reflecting the third applicant's natural family ties amounts to a failure to respect her family life. Moreover, the close and intimate relationship between the third applicant and her parents is such that there is of necessity also a resultant failure to respect the family life of each of the latter.'

The next morning the *Irish Times* editorial headline was simply 'Children's Rights':

> This is a state which loudly professes its love of children and concern for their welfare. The fact that we have no divorce and no right to remarriage means that many more children than would normally be the case find themselves born 'illegitimate'. Illegitimacy is, and always has been, a stigma. And the European Court of Human Rights has now ruled that such children are being denied fundamental rights enjoyed by 'legitimate' children.
>
> It is embarrassing enough that it should take a decision of the court to force us to treat all our children equally; the ultimate ignominy would be if we were to delay or procrastinate any further about it.

The Johnstons' very considerable legacy was a change in the law to remove the distinction of 'illegitimacy' (effected by the Status of Children Act 1987) and the lifting of the taboo of families living outside wedlock, the notion of 'living in sin'.

In 1983 the post of attorney general had become vacant when the incumbent, Peter Sutherland of Fine Gael, was appointed as the Irish commissioner to the European Community. The Labour Party, the junior party of the then-coalition government, was enabled to choose the new attorney general as part of the overall political balance. I was keen to serve and felt I had the skills, experience, and independence for the post. Dick Spring, from Tralee in County Kerry, was now the leader of the Labour Party and Tánaiste (deputy prime minister). I told him that I would like to serve as attorney general, and made my case as to why I felt I was a good candidate for the job, emphasising my independent disposition, which I felt was a plus.

The post would have meant joining the cabinet and really being part of Labour in government, much more than as a backbench senator. As the first woman to serve as chief legal officer of the state, it would also have broken a glass ceiling. But it wasn't to be. John Rogers, a junior counsel who needed swiftly to take silk to accept the appointment, was given the job. John and I were friendly if not particularly close at the time, and I recognised that he was bright and talented. A good friend of Dick Spring's and very much a party backroom person, John was a good team player, and I had clearly over-emphasised my independent disposition.

My pride was hurt, and though I tried to hide it, Nick could tell that it had come as a blow. In the event, John proved himself a worthy attorney general, and I found solace in the cases I was working on.

One of these cases, *Norris v Ireland*, concerned gay rights. When David Norris, a lecturer at Trinity College Dublin, and a peer of mine approached me about taking a case against Ireland on his behalf, I asked him to explain to me what exactly it had been like for him growing up in Ireland. He wrote me a long, very moving handwritten letter, and by the time I had finished reading it I had a much deeper understanding of his plight and of the importance of taking his case.

About the age of 13/14 I first noticed that feelings which were quite natural and spontaneous to me were not shared by my closest companions. When I broached the subject it led to a number of heated arguments during the course of which I discovered to my horror that other people did not share my ability to respond to my own sex, and that these feelings which I had regarded as part of everyone's experience were condemned and regarded as criminal.

No matter how subtle or convincing the defence of my position, we always came back in discussion to the same two points: that my nature was sinful and criminal.

As far as the question of sin was concerned, I was, and am a practising member of the Anglican Communion, a church that allows a
certain latitude for individual conscience in the relationship between
the individual and God, so that while I knew that a number of the
clergy despised people like myself, I was confident that my nature
was not evil in the sight of the God who had created it.

However, the idea of being a criminal without even knowing it
was much more disturbing as it was not a matter of opinion but a
political fact. If I expressed myself as my instincts increasingly urged
me to, I became de facto apparently a member of the criminal classes
liable to arbitrary punishment and disgrace. As I was of a literary turn
of mind even then my attention was jocularly drawn to the facts of
another Irishman – Oscar Wilde. But it was no joke for me, and I was
shocked to discover that the very same laws under which he had
suffered so many years before still applied in Ireland.*

Having suffered loneliness and depression during his youth as he tried
to come to terms with his nature, living in fear of criminal sanction or
exposure through blackmail or loss of his employment, frustrated by
the lack of positive information or any kind of support for homosexuals in Ireland, David Norris courageously came out publicly, and
founded and presided over the Irish Gay Rights Movement.

In the late 1970s he decided to challenge those laws criminalising
certain homosexual activities as being unconstitutional and an interference with his right to privacy. 'I do not ask for mercy, compassion,
pity or well-intentioned advice,' he wrote, 'but for the justice which
I believe is enshrined as my right and entitlement as a citizen under
the constitution of this land. I seek the vindication under law of my
right not to be unthinkingly and unfeelingly rejected by my native
land, and of my right to conduct my personal relationships with
dignity and decency in accord with my nature, and without the
threat of police intervention or the possibility of blackmail.'

I took his case, starting in the Irish High Court in 1977. The laws
in question dated back to the Victorian era: the 1861 Offences Against

---

* Reproduced with the kind permission of Senator David Norris.

the Person Act and the 1885 Criminal Law Amendment Act. We argued that because the relevant sections were repugnant to the Irish Constitution of 1937, these laws were not continued in force at the time of the Constitution's enactment and therefore did not form part of Irish law. We sought a declaration to that effect. Notwithstanding its finding that 'one of the effects of criminal sanctions against homosexual acts is to reinforce the misapprehension and general prejudice of the public and increase the anxiety and guilt feelings of homosexuals', the High Court refused David Norris's application, upheld the constitutionality of the laws, and dismissed the case.

On our appeal, in 1983, that judgment was affirmed, the Supreme Court considering the laws making homosexual conduct criminal to be consistent with the Irish Constitution and holding that no right of privacy encompassing consensual homosexual activity could be derived from 'the Christian and democratic nature of the Irish State'. The majority judgment stated, 'Homosexuality has always been condemned in Christian teaching as being morally wrong. It has equally been regarded by society for many centuries as an offence against nature and a very serious crime.'

The Court did, however, award David his legal costs of both actions, recognising that he had raised issues of exceptional public interest. He had also, by standing up for his rights and revealing details of his personal life, brought the whole issue into the open, provoking a national debate and a huge amount of press coverage. And, indeed, a minority judgment delivered by Mr Justice Seamus Henchy had recognised that '. . . fear of prosecution or of social obloquy has restricted [Mr Norris] in his social and other relations with male colleagues and friends; and in a number of subtle but insidiously intrusive and wounding ways he has been restricted in or thwarted from engaging in activities which heterosexuals take for granted as aspects of the necessary expression of their human personality and as ordinary incidents of their citizenship.'

David was not about to give up on this long crusade; there was too much at stake, not just for him personally but for the entire gay community in Ireland. We had one further forum to which we could appeal: the European Court of Human Rights in Strasbourg.

Before the Strasbourg court, our argument was slightly different. As a contracting party to the European Convention on Human Rights, Ireland had signed up to Article 8 of the Convention, which provided that everyone had the right to respect for his private and family life, and that that right could not be interfered with by the state except for a legitimate aim, such as for the protection of health and morals or a pressing social need.

David Norris had been elected to the Senate in 1987, and one of the Irish Government's arguments was that he had not been prosecuted under the impugned statutes and had been able to maintain an active public life side by side with a private life free from any interference on the part of the state. The government further argued that contracting states enjoy a 'discretion in the field of the protection of morals', meaning this was a matter for Ireland's institutions and laws, not Europe's. The Court disagreed, holding, rather, that the case concerned a most intimate aspect of private life and that there had to be particularly serious reasons before state interference could be deemed legitimate. The court could see no such particular justification for retaining the laws, and therefore ruled, in a judgment delivered in October 1988, that Ireland had breached Article 8.

'Sound judgment,' read the *Irish Times* editorial the following day. 'Irish legislation on homosexuality has long been out of step with Europe and [with] enlightened modern thinking. The time has come for change; it is a sad state of affairs that, once again, it has to be forced on us [by the European Court].'

Finally, after more than a decade, Senator Norris had won his case and forced the Irish Government to repeal the law. I was delighted for David, and for all those who had been living in the shadows who might now feel more secure in their identity. I had identified this issue in that undergraduate speech on law and morality I had given some twenty years earlier. It was a sweet victory to savour.

In 1981, my brother Aubrey, in his late thirties, was diagnosed with cancer. He was badly stricken and went through brutal chemotherapy. He was in hospital in Dublin for that period, and I went and sat with him every evening. I was pregnant, and when I had a boy, there was, of course, only one name possible. Aubrey struggled hard; he was in terrible pain; it was hard to comfort him. Then he gradually got better. He had a good remission.

About two years later a second wave came. Aubrey was drinking heavily, trying to rise above it, but he was a doctor, and he knew the score. With more chemotherapy he got a slight remission over Christmas 1983, which we all spent together with my father in Ballina. On New Year's Day 1984, Oliver drew Henry, Adrian, and I aside and told us that we had to face up to the fact that Aubrey was not going to last much longer. I remember looking at Henry and Adrian blankly, the three of us thinking, *this is unbelievable, this cannot be true.* We just couldn't fathom that Aubrey was going to die. I have compared notes with them since, and none of us like New Year's Day. It absolutely brings back those feelings of anguish, helplessness.

The weeks went on, and we tried to prepare ourselves, tried to face the reality that Aubrey was not going to get better. Now it was a question of quality of life. His wife, Pamela, showed great inner strength. It broke my heart to see her with their three boys, the youngest, Charles, the same age as my little Aubrey, just three. Aubrey himself was reconciled by now, and extraordinary about it, using his humour and charisma to cheer us all up.

Then, just before St Patrick's Day, Nick and I were in Boston at a conference to celebrate 'Irishness'. Seamus Heaney, whom I had remained friendly with, was also there, and we had some good discussions on the theme. But we got word to come home at once: Aubrey had taken a sudden turn for the worse, and there was not much time. We scrambled onto the next flight home, to Shannon Airport, and drove the two hours to Castlebar Hospital. For the whole journey all I could think was, *I can't believe this, I can't believe Aubrey is going to die.*

Aubrey was emaciated from the cancer at this stage. He knew us all. We were all in and around the bed, coming in and out, talking to him, trying to comfort each other; and he was cheerful, but fading fast. We gathered around. The time came when he could no longer talk. His breathing became shallower. In a final, beautiful gesture he put his hand to his mouth and, one by one, blew us a kiss. Then he lost consciousness for good and slipped away from us.

Of course the funeral in Ballina was massive. The funeral Mass and procession to the graveyard helped us all to cope, supporting one another. But I have never felt such grief. It kept coming and coming; I was inconsolable. At the same time, watching Aubrey die had taken away the fear of death. It was as if he wanted to offer one last gift by the manner of his dying. Was he in heaven? I wanted to believe so. My father was consoled by his deep faith, and I found comfort myself in the familiar prayers while continuing my private journey to find answers to unanswerable questions. Aubrey's death seemed so much crueller than my mother's. He was too young to die, so close in age to me that we had had an almost telepathic empathy. With his loss, somewhere inside me a light went out.

During the 1980s a terrible backdrop to Irish life, public and private, was the ongoing sectarian conflict in Northern Ireland. At issue was the constitutional status of Northern Ireland, and the relationship between the majority Protestant unionist community, generally in favour of British rule, and the minority Catholic nationalist community, generally in favour of a united Ireland. The conflict, known as the Troubles, had started in the 1960s with the formation of a Protestant paramilitary force, the Ulster Volunteer Force (UVF), and with attacks on a civil rights march in Derry, which led to a resurgence of the Catholic paramilitary organisation, the Irish Republican Army (IRA). British troops were deployed to Northern Ireland in 1969. Atrocities were committed by all sides, and by the 1980s the politics of violent force were overshadowing the democratic parties that rejected violence.

In 1983 Taoiseach Garret FitzGerald established, for the purpose of political discussions designed to alleviate the conflict, the New Ireland Forum (of which I was an alternate Labour Party member) with membership from the main political parties in the Republic and from the Social Democratic and Labour Party (SDLP), the nationalist party of Northern Ireland led by John Hume, which rejected violence and represented democratic constitutional nationalism. A New Ireland Forum report published in May 1984 proposed three possible alternative structures: a unitary state (i.e., a united Ireland); a federal/confederal state of Northern Ireland and the Republic; or a joint authority, meaning joint Irish- and British-Government responsibility for the administration of Northern Ireland.

In November 1984 british prime minister Margaret Thatcher, following an Anglo-Irish summit with FitzGerald, made what became known as her 'out out out' speech in a forceful rejection of these three proposals on Northern Ireland.

Shortly after that, Nick and I travelled to the Aspen Wye River Conference Center, near Washington DC, for a working weekend conference of politicians and academics to discuss Northern Ireland matters with some of the central players. Members of the SDLP were demoralised and in disarray. In sharp contrast the contingent from the unionist parties, including Peter Robinson and Bob McCartney QC, were cock-a-hoop and exuding confidence.

This was a time of terribly low morale from the nationalist perspective and from the Irish government's perspective as well. I sensed a real desperation on the Irish side to get some agreement with the British Government on the Northern Ireland problem.

A year later, with Garret anxious to get an agreement and Mrs Thatcher facing an election at home and therefore under pressure, the two governments signed the Anglo-Irish Agreement. Mrs Thatcher conceded to an agreement that would enable a British-Irish joint executive, the first hint of an alternative to, or a dent in the idea of, British rule in Northern Ireland. In the end, the joint executive never came about because the unionists were so opposed. Outraged, they felt they had been sold out by Britain and betrayed by Margaret Thatcher. A week after the agreement was signed, a mass rally outside Belfast City Hall was attended by tens of thousands, perhaps a hundred thousand, from the unionist community, cheering to the echo as the Reverend Ian Paisley, leader of the Democractic Unionist Party, proclaimed, 'Ulster says Never!'

I was concerned that the Anglo-Irish Agreement would never be supported by unionists, who felt it had been imposed on the unionist majority without adequate consultation. Its stated objective was to secure peace and stability within Northern Ireland, and I doubted it would achieve this. Frank Cluskey, also a Labour Party member of the New Ireland Forum, expressed similar concern, as did Paddy Harte of Fine Gael. The general mood, however, was of celebration; there was joy beyond belief in Dublin that an agreement of any kind had been achieved after that terrible low.

I felt acutely the mood of triumph in Dublin, and how hurtful it was to many in Northern Ireland. As a senator representing many graduates who lived in Northern Ireland, I felt I needed to be free to express my concerns publicly, and to work for a better Anglo-Irish agreement. I had already voiced these concerns to the Labour Party leadership before the agreement was signed, and was dismayed when a communiqué was issued stating that support for it was unanimous. I knew I had to act. I contacted a few close allies, sought their advice and came to my decision, announcing that I was resigning from the Labour Party in order to go back to the independent benches and speak freely.

In the Senate, rather than vote against the agreement, I absented myself from the chamber, thereby abstaining. I met Garret FitzGerald privately the day after my resignation from Labour and my public criticism of the agreement, and we had a healthy discussion about the whole thing. He recognised that I felt passionately about it and was acting in good faith, and made his case about the positive aspects of the agreement as he saw them. There was no question of our falling out over it, and that was important to me, personally. And there was no terrible falling out with the Labour Party leadership, either.

In retrospect – and it was not evident to me at the time – what the Anglo-Irish Agreement did was break the mould and start a whole new debate, within broader parameters, where the Irish and British Governments worked more closely together in trying to solve the Northern Ireland problem. This led, eventually, to the Peace Process and the Good Friday Agreement in 1998. What Garret saw that I didn't see then was that persuading Margaret Thatcher to enter into an agreement with the Irish Government on Northern

Ireland meant an irrevocable mental and political shift on the part of both governments; now they had a shared objective. The Anglo-Irish Agreement itself did not work, but it began a process that would ultimately be successful.

In the general election of 1987, I stood for a Senate seat, campaigning as an independent again, and secured it fairly comfortably. But when the following election was called two years later, in the summer of 1989, I decided not to run. I had been a senator for twenty years, but I no longer had the same driving energy to continue to contribute through that forum. It was not a great time to be introducing legislation under a Fianna Fáil-led government, of which Charlie Haughey was now Taoiseach. The legal cases I was involved in were having more impact and enabling me to bring about change. If I feel that I am not giving of myself fully, I have to stop. So I decided not to contest again – and by not putting my name forward, I effectively retired. In essence I had come to the conclusion that I'd had a good run but that my legal work would be my main focus from here on in. I hadn't exactly excelled as a party politician, and I had perhaps achieved what I could as an independent senator. Now it was time to move on. To some extent I missed the cut and thrust of parliamentary politics, but being on my feet arguing cases presented plenty of challenge.

Nick and I had established the Irish Centre for European Law in 1988, and this occupied a large proportion of our time, energy, and enthusiasm. The centre was needed, we believed, because more and more European legislation was having direct impacts on a wide range of areas in Irish law: agriculture, competition, corporate law, environment, equality issues, labour standards. Our aim was to provide an independent centre where we could bring together expertise and present and publish lectures of a high quality on relevant subjects, to meet the significant demand for continuing professional education on European directives and European law.

Nick was enthusiastic about the potential role of the Irish Centre for European Law and became its administrator, initially on a voluntary basis, and later on a part-time salary. We worked well together, having complementary skills. Nick brought good practical administrative and organisation skills to the centre's work and I focused on deciding on the issues of European law we would examine, inviting the experts, and moderating the discussions. Nick would produce brochures advertising the conferences to our members, drawn from different sections of Irish law. We recorded the sessions, and Nick took responsibility for editing the published reports, which were much in demand.

We had wanted the centre to be independent but located in Trinity, so we approached the provost of Trinity, William Watts, and also the law school, and they immediately got behind the idea. They agreed to host the centre and be a venue for the lecture series, but understood that the centre was to be independent and broad-reaching. Tom O'Higgins, former Chief Justice and then a judge of the European Court of Justice in Luxembourg, agreed to be our first chair, bringing his unique combination of good humour and gravitas. We put in a huge amount of work and were gratified by the high response to the conferences, which filled the large lecture halls in Trinity and ventured out to the law schools in Cork and Galway. Nick and I saw this as a long-term, absorbing project.

I had also joined chambers in London, at 2 Hare Court, around 1987, at the invitation of my friend Anthony Lester QC, one of the great human rights lawyers of our generation. This was a big step in my legal career. As a sole practitioner in Ireland with some experience arguing before the European courts, I had a reasonable chance of getting briefed in cases (if any turned up) that might go to Europe. But by joining these prestigious chambers, which operated under a very different system from the central library one in Ireland, I would have the opportunity to take on many more European briefs. So now I was travelling regularly to London for meetings and consultations at chambers, and to give talks on my specialised subjects.

On the morning of 14 February 1990, John Rogers, the former attorney general, with whom I was quite friendly, telephoned me asking if he might see me at home as he had something he wanted to discuss.

Presuming that he had a personal or legal problem on which he wanted my advice, I told him to come on over. The children had gone to school, and Nick had gone to the centre in Trinity. When John arrived I brought him in to the dining room and asked him what I could do for him. He said he'd been authorised by the Labour Party to say that they would like me to consider becoming a candidate in the forthcoming presidential election. Dick Spring had, a few months earlier, declared that Labour would be contesting the presidential election (as, for the first time, Labour had a sufficient number of parliamentary members to nominate its own presidential candidate) and they were developing ideas on 'a working presidency', more relevant to the Irish people. At the suggestion of Ruairí Quinn's secretary Denise Rogers (no relation to John Rogers), they had decided to ask me, as they felt that, with my track record, I was someone who could make a difference. I seemed to fit Labour's 'job description' for this role of working president: as a former senator and constitutional lawyer, I could be relied on to understand both the limits and the potential of the office. It also helped that I was a woman, and, in my mid-forties, relatively young.

I looked at John in absolute astonishment, my jaw dropped, my eyes widened, and I immediately said to myself, *no way!* I had a good life; it was good for Nick; it was good for the children. We had the centre; we had my work in Ireland and Europe; I had important cases coming up, and life was exciting. As far as I was then concerned, the president was very much the ceremonial head of state: style and little substance, receiving ambassadors, red carpet events and all that, but no role that I could envisage for myself.

I tried to hide my feelings from John (and did so so poorly that he told me afterwards he was sure I was going to say no). I felt that it was an honour to be asked and that I should take the time to consider my answer. The presidency, after all, is the highest office in the land. So I said that I would think about it and let him know the following Monday. As I let John out the door, I was still shaking my head in disbelief.

I telephoned Nick. I had to go into court that morning; Nick was already in Trinity. We hadn't planned to meet that day, but when I called him he said, 'Come to lunch. It's Valentine's Day!' Over lunch, Nick urged me, 'Look, Mary, you should think about this. The very least you might do is look at the Constitution.' I felt a little spark of adrenaline. When I got back to my office I took out my well-thumbed copy of *Bunreacht na hÉireann*, the small blue book of the Irish Constitution, with the gold harp on the cover. I opened it at the relevant article and read the oath:

> In the presence of Almighty God I do solemnly and sincerely prom-ise and declare that I will maintain the Constitution of Ireland and uphold its laws, that I will fulfil my duties faithfully and conscien-tiously in accordance with the Constitution and the law, and that I will dedicate my abilities to the service and welfare of the People of Ireland.

I began to think. This was a president elected directly by the people; that had a significant impact. I slipped the copy of the Constitution into my handbag and pulled it out and read it and reread it over the weekend.

I was due in London that weekend for a meeting at chambers with Anthony Lester QC. I had told nobody, except Nick, about my conversation with John Rogers. I decided to confide in Anthony because he was my friend, and he was discreet. Using a slightly self-mocking tone, for whatever reason, I said to him, 'By the way, you might be interested to know that we're going to have a presidential election in October this year, and I've been approached by the Labour Party to be their candidate.' Anthony looked worried, even aghast. He implored me, 'No, no, Mary, please don't accept, because if you run you will win!' I laughed in his face and replied, 'Oh, Anthony, that just shows how little you know about Irish politics.'

9
A President With a Purpose

I am of Ireland. Come dance with me in Ireland.
WB Yeats, after a fourteenth-century Irish poet

ALL THAT weekend my mind was whirring. I telephoned Bride
Rosney from London, asking her if she would collect me at the
airport as I wanted to run something past her. As we drove in towards
the city centre, I dropped my bombshell, telling her about my
conversation with John Rogers. Bride was so surprised that she had
to pull the car to the side of the road. We sat in silence for a moment.
I took the Constitution from my handbag, and read out the oath of
office: '. . . I will dedicate my abilities to the service and welfare of
the People of Ireland.' We talked about how you could be creative
with that; it was such a personal statement. Because the candidate
had been unopposed on the previous three occasions, there hadn't
been a presidential election for seventeen years. It was generally
assumed that any contest would be won by the expected Fianna Fáil
candidate, the popular deputy prime minister, or Tánaiste, Brian
Lenihan. But if I entered the race, I could contribute to a new idea
of the Irish presidency, an office of head of state with little real
executive power but with scope for influence, moral influence, and
the potential to define the modern Ireland. My goal would be to
start a debate on expanding the role of the president, but I saw this
as something the eventual winner, presumably Brian Lenihan, would
run with.

That Sunday evening, Nick and I sat down to consider the implications for us as a family. I was excited at the prospect of running,
but still had no thought of actually winning. What about the children? Tessa was in her final year of school and facing her Leaving
Cert. exams, but they would take place in early summer, before the

campaign became too all-absorbing. (The election would be held in November.) William was in his fifth year in secondary school, doing well, and Aubrey, then eight, was in primary school just around the corner from our home. If I was on the road campaigning, my legal practice would be neglected for those months, but Nick reassured me that I would be able to pick it up again in the autumn, and that we would be able to cope financially until then – and wasn't this just the kind of challenge I was made for? So we decided to let events take their course.

When we broke the news to our children that I was considering running for president, they took it well. They were used to seeing me on the news or on political programmes on television from time to time, and seeing my picture in the papers, and they had some sense of what a political campaign is like, having helped over the years stuffing Senate election brochures into envelopes. They didn't particularly relish the idea of the extra attention they would get from friends at school, but they were as infected as Nick was by my renewed energy and sense of purpose.

I never explicitly or formally asked Bride to help me with the campaign, as it was to become; rather, her help was simply assumed on both sides. From then on she was always nearby; apart from Nick, she was the first person I consulted on any issue.

It was clear that I should not rejoin the Labour Party. Having resigned a few years earlier on a matter of principle, it would be inappropriate to return in order to be nominated as candidate for a post perceived to be above politics. If I wanted to run – and Bride could also see that my adrenaline was up – I should run as an independent nominated by Labour. One of the perceived limitations of the presidency was precisely this idea that it was above politics. We could turn that into a virtue. 'But,' Bride warned, 'the Labour Party will expect you to rejoin. They may withdraw the offer if you refuse.'

I phoned John Rogers that Monday and, thanking him, said that I would be prepared to run but explained why it would have to be as an independent candidate nominated by Labour. John indicated that this was probably not what Dick Spring had in mind, but that he would get back to me. *So be it*, I thought, *that's the end of that.* But I felt a twinge of disappointment. I saw the challenge of arguing for a more active presidency as akin to a legal brief. My attitude with John was, well, do you want me to take on this brief or not?

There followed a discussion within the Labour Party ranks. Noel Browne was the preferred candidate of some, including prominent TDs (members of parliament) Michael D. Higgins★ and Emmet Stagg. Speculation began in the newspapers. A few weeks later, John Rogers came back to me and said that my nomination would be put before the party's administrative council. On 26 April 1990, by a comfortable vote, they accepted me as candidate. What a rush! My energy was sky high, and I realised then just how much I wanted to go out and campaign for this, how much of an opportunity it was to contribute to a debate about shaping a modern Irish identity.

I had a meeting with Dick Spring and some of his advisers, including John Rogers, and Dick assured me that I had the full support of the Labour Party. Nick and Bride were at that meeting with me, and we raised the idea that we should now seek the support of other, smaller parties, such as the Workers' Party and Democratic Left, and some like-minded independents. This was not greeted with huge enthusiasm by Dick Spring or Labour, and indeed rows between members of these various parties would become a feature of the campaign.

The election would not be held until November, so this would be, by Irish standards, an extremely long campaign. I received invitations to speak in towns and communities around the country, beginning a journey that changed and enriched me. I saw the beauty of the landscape and the calibre of the people in places such as Allihies, where I attended a conference about communities on the periphery coming together in mutual support. Their conviction that when you are on the periphery the periphery is the centre was an early lesson in the importance of small places.

On 1 May the campaign was formally launched in Limerick city. Jim Kemmy, a disaffiliated Labour TD from Limerick who had

---

★   Michael D. Higgins became the ninth president of Ireland in 2011.

established his own party, the Democratic Socialist Party, took the lead on this launch with his colleague Jan O'Sullivan. I had long admired Jim for his honest politics and his sense of equality and justice, his great warmth and humanity. A stone mason with burly shoulders and a broad chest, he was recognised as an authority on the history, buildings, and culture of Limerick, a city and county he loved with a passion.

The launch went well and the national and local press were there to cover it. At one stage as I stepped aside and quietly powdered my nose, Jim decided to give me some advice. 'Mary, if you're going to do this seriously, you're going to have to tidy yourself up. You need to get yourself a makeover.' Part of me was laughing at the idea that Jim, in his rumpled suit, was counselling *me* on matters of sartorial elegance; and part of me realised that I must indeed have an image problem: my hair needed styling, I wore minimal make-up, and my clothes were dark and lawyerly. Against my natural inclination not to bother much with how I looked, I would have to take Jim seriously.

Next morning, Jim and I did a live interview on a local radio station. After a few questions and some discussion, the interviewer put it to us that there had been criticism of the current president, Patrick Hillery. What did we think of him? As one of our campaign objectives was to advocate for a more active presidency, expanding the role, I was tempted to make that point, implicitly criticising Hillery's presidency. But before I could say anything, Jim Kemmy jumped in and answered. 'President Hillery is a very fine man and has been a good president for this country,' he said emphatically, catching my eye as he continued: 'He understands the office, he has been independent in the way he exercised his functions . . .' Jim Kemmy taught me an important lesson: to respect the office and the office-holder, never to speak ill of your predecessor. First of all, Jim was absolutely right. Paddy Hillery had been a good and decent president under difficult circumstances. It is a lonely office but, in the best interests of his country, he had agreed to serve a second term, which he had not wanted to do. Second, I learned that I needed to conduct my campaign in the tone of somebody who was seeking high office, who understood the nature of that office, rather than focusing on personalities in a partisan way.

Meanwhile, I had the challenge of overhauling my image, which did not come naturally to me. It had been many years since I'd had my mother's invaluable help with styling. At a gathering of business-women, a Dublin socialite had said dismissively, in my hearing, 'She doesn't have the wardrobe for it.' Initially my innocence was such that I wondered what on earth you would need a wardrobe for, thinking, literally, of the piece of furniture. But I accepted the point, of course: running for the presidency was effectively going for a job interview; the interview board was the electorate, the people of Ireland. Everybody going for an interview puts on her best bib and tucker and gets her hair done. As a woman, if I was going to gain the confidence of the people, objectively and professionally, I had to look good. I had to look the part.

I was fortunate to be offered the services of public relations experts, including Brenda O'Hanlon. Brenda was a breath of fresh air: professional, competent, full of energy. She had a good eye for clothes and brought in Cecily MacMenamin of Brown Thomas to put together a 'wardrobe' for me. I was beginning to get it. I was given a makeover and, for the first time, got my hair done regularly and applied full make-up. Brenda organised a colour article in the popular magazine, *Hello!*; a magazine, I have to admit, I had never heard of, but that was reputedly read in every hairdresser's and every doctor's waiting room in the country. The article had quite an impact; it generated more interest in the mainstream media. As I travelled around the country, I gave lengthy interviews to local radio and regional newspapers that included photographs of me with local people – now looking more like a presidential candidate.

When my name had first been bandied about as a candidate, Nick and I received, unsolicited, an intriguing letter from Eoghan Harris advocating a fresh and unorthodox way in which the campaign should be run. I knew Eoghan's name but little about him, other than that he had worked at RTÉ, the national broadcaster, and had recently been connected with the Workers' Party as a media adviser, although the relationship had broken down. Eoghan's letter struck a chord with Nick and me. He set out his stall, and outlined a vision that melded down-to-earth common sense and inspiration. Importantly, even at this early stage, he was of the opinion that I could win the campaign – or come close enough to call it a famous moral victory – but to achieve this, I would have to broaden my reach beyond the intellectual, liberal space I then occupied.

Right now you have a whole country to win. This does not mean compromising your principles. But it does mean accepting that every morning we invent a new world. In politics you need to put previous political personae aside and start afresh. People like fresh starts.

We took Eoghan up on his offer of practical help, and he treated us to a stunning intellectual analysis of how to run the campaign. If I talked about nothing but liberal issues then I would be giving people a narrow image repeating what they were expecting, given my background. Whereas, if I wanted to be a president for all the people, I had to appreciate the broader dimensions of what was happening in Ireland, and the wider sense of ourselves that needed to be captured. You have to have seen Eoghan Harris in full flow to appreciate the experience: the breakneck pace, the wit, the revelatory insights that had us saying, 'Ah, yes, of course.' There was genius in it.

I was not by any means a household name when I was nominated – the Irish bookmakers rated my chances as 100 to 1 against. According to Eoghan, I needed to move from my 'plummy Dublin lawyerly image' to 'a warmer image; the common touch.' He advocated that I enjoy every moment of it, that a spirit of adventure and zest, and sheer energy, would enthuse the electorate. And I did relish it; it was not insincere. I felt it was absolutely the right approach. It tied in with a strong theme that arose during the course of the campaign, the 'spirit of *meitheal*'.

'*Meitheal*' (pronounced with a silent '*t*') is an Irish word that describes a traditional, rural practice of people coming together to work, farmers lending support to their neighbours as the need arises.

It expresses the idea of community spirit and self-reliance. I remember as a child going out with my father on his calls to rural areas at harvest time: practically everyone would be working in a particular field to save the hay, the women bringing sweet tea and bread and jam. If a farmer was sick, his field would be harvested willingly by neighbours. I find *meitheal* similar to the idea of *ubuntu*, an African ethic that I have become more familiar with since, an idea of human interconnectedness and solidarity, described by Archbishop Desmond Tutu when he says, 'I am because you are.'

In the summer months, before the two big political parties had even declared a candidate – though everyone knew Fianna Fáil was going to put up the popular Brian Lenihan, a witty and astute political veteran who had fought bravely against illness, and everyone was sure he would be Ireland's next president – as I travelled all over the country to canvass, time and again people raised the idea of *meitheal*.

In 1990, Ireland was still a relatively poor country when compared with its European neighbours, but local, voluntary self-development was evident everywhere, through GAA★ clubs, residents' associations, women's organisations, and community volunteers. With our membership in the European Union, and the common agricultural policy, more money was coming in to rural Ireland, to towns and parishes, but not enough for these areas to have the facilities and opportunities the cities had. So people were banding together to build centres for young people, sports facilities, space for the elderly, theatres, community halls. I tried to capture this. I promoted the idea that the president could be much more proactive, much more in touch with people, could do more in fulfilling the oath. There was nothing in the Constitution to prevent this. The president has express constitutional powers: to refer bills to the Supreme Court to test their constitutionality, to address the Oireachtas (the houses of parliament), to address the Irish people in a time of national emergency, and to fulfil diplomatic duties. But as a constitutional lawyer, I grasped the significance of the fact that the people had a right to elect their head of state directly in a nationwide poll, which not

---

★   Gaelic Athletic Association.

many countries allow – elect someone who was part of the Oireachtas but also outside politics, someone who could realise the full potential and dignity of the office, and shape a modern sense of Irish identity.

There was a feel-good factor in the air. The Berlin Wall had fallen in the autumn of 1989. The Velvet Revolution had taken place in Czechoslovakia. And that summer of 1990, the Irish soccer team had for the first time qualified for the World Cup finals, taking place in Italy. Italia 90 had an amazingly positive impact in Ireland, and Eoghan Harris captured this in a dazzling television broadcast showing images of me in a market chatting and laughing with people, to the uplifting strains of the Italia 90 theme, Puccini's 'Nessun Dorma'.

The campaign evolved organically and divided into different natural layers. We formed an official campaign committee with the Labour Party, chaired by Ruairí Quinn, a long-time friend who was a fair-minded and skilful director of elections. The official committee included the secretary of the Labour Party Ray Kavanagh; John Rogers; Fergus Finlay, Dick Spring's special adviser and a renowned strategist; and 'the Robinson camp': Nick, Bride, me, and Peter MacMenamin, a Labour member and trade unionist, and a close friend of Bride's. There was a wider political grouping with the Workers' Party and Democratic Left. In different constituencies there were fierce fights between the political parties as to who would be on the platform with me, and back channels had to be opened by Bride to keep people informed. We worked hard to get the smaller parties involved in that committee, often against local Labour wishes, and we eventually brought in Workers' Party representative Eamon Gilmore (as I write, Tánaiste and leader of the Labour Party).

Eoghan Harris had offered his services privately, aware that the Labour Party might try to veto his involvement. When his role was eventually revealed, Labour was understandably furious. I regretted the subterfuge involved, which was quite out of character, but by this stage we had made such progress that wiser counsels prevailed and they coexisted uneasily thereafter.

My brothers Henry and Adrian, of course, became involved in the campaign, particularly in their own localities, though both kept their distance from the political parties. Henry was now in Galway, and Adrian was back in Ballina. (My other brother, Oliver, had emigrated to New Zealand with his family.) And a posse of volunteers, not particularly politically aligned, were supporting me as an independent candidate, some with their own high profile.

It was encouraging that as I went to events (sporting fixtures, the national ploughing championships) and visited different parts of the country, people began to queue up to meet me and shake my hand, so that my right hand was quite bruised by the time the election came around. The challenge brought out the part of me that is more like my mother – listening to people, engaging, hearing their ideas.

I had performed respectably in the opinion polls from the beginning. I can remember one Sunday morning early on, before the other candidates had declared, Nick reading out a poll in a Sunday newspaper with a list of speculative names. I polled at slightly over 20 per cent. We were pleased with that figure, excited even. As to first-preference votes, there was no way I could out-poll Fianna Fáil's popular Brian Lenihan. Our target was to beat the Fine Gael candidate to second place, then take sufficient transfers from that eliminated candidate to challenge for the lead. But Fine Gael, the second-largest party, was having difficulties securing a candidate. Our worst scenario would be if Fine Gael decided not to run a candidate but instead rowed in behind me. For me to come close, it had to be a three-horse race.

In the end, Fine Gael did field a candidate, Austin Currie, a northern civil rights activist of the SDLP who had joined Fine Gael. When they were nominated, both he and Brian Lenihan began to embrace the idea of a more proactive presidency that I had been advocating, but they seemed less sure-footed, less convincing.

Around the end of August or the beginning of September a private poll indicated that if I polled ahead of the Fine Gael candidate, I would win. I remember feeling quite shaken, and a stream of thoughts bombarded me. What does this mean for me? What does it mean for Nick and the children? What will we do with the house? What about my work? My cases? And another thought sneaked in: do we really want to do this? Then the adrenaline took over. Okay, if this is winnable, I am going to go and win it. I changed my mental approach and made a conscious decision then, to fight to win.

My campaign slogan was 'A president with a purpose'. Some elements of the media kept repeating that the job of president was a non-job, one that involved carrying out some diplomatic functions as a figurehead, inspecting guards of honour, and inviting dignitaries up to Áras an Uachtaráin (the president's formal residence) for tea. But I was advocating a proactive presidency founded on inclusivity, the idea of a president for *all* the people. I was convinced it was a big job, and needed subtlety. The president is the only person in Ireland elected individually to office. In a general election, when the results are declared, one group of individuals belonging to the largest political party, or a coalition, takes power and appoints people to cabinet and other roles. But in presidential election, a person is elected by the people to represent them as their head of state, and to fulfil the oath of office. That role of representing the nation – outside politics – impacts many aspects of Irish lives: how we perceive ourselves at home and how we project ourselves abroad; how, for example, we link as Irish people with the people of Northern Ireland, with those who have left Ireland; how we link with developing countries and deepen our engagement with them.

Despite my commitment to the campaign, and the fact that I was in it to win, against everybody's advice I took time out briefly very late on, flying to Luxembourg to argue a case before the European Court of Justice. Bride was furious, understandably, and even Nick was questioning my judgment; it was sending out the wrong message to the electorate; they argued, if I was keeping up my old job, still arguing cases. But I was adamant. I felt I couldn't abandon my responsibility as counsel to my clients.

The case, *Cotter and McDermott v. Minister for Social Welfare*, had quite a long history, and I had been intimately involved in it from the start. It involved a claim of discrimination against married women under the social welfare code. In essence, married men were the automatic recipients of state child benefit payments, and received higher rates of benefit payments on the assumption that their wives were dependent on them. For a married woman to receive the higher rate of payment, she had to prove that her husband could not support himself. In 1984 a European directive required Ireland to stop this discriminatory practice. Ireland didn't implement the directive until some two years later, and after expiry of the period allowed. On an earlier reference to the European Court, we had successfully argued that the women's right to payments at the same rate as married men commenced in 1984, as a result of the directive in Ireland, rather than in 1986, when the Irish law was changed. Now the High Court was refusing to apply the law retroactively, to compensate for past discrimination as it were. So the Supreme Court had again referred the case to Luxembourg, this time on the specific issue of retroactive applicability.

When the president of the court was indicating the time frame for a judgment, he smiled at me as he said, 'Mrs Robinson, the court is aware that you may or may not be present to receive the judgment.' Happily, then, I was not present in Luxembourg when the judgment went entirely in favour of our clients. When applied by the Supreme Court, that decision cost the state some £200 million in back pay to married Irish women who had been discriminated against.

In early October 1990, the formal nominations took place in the Custom House in Dublin, and we decided to launch the final drive of my campaign in Ballina, rather than Dublin, the predictable venue. Henry, Adrian, and my father had built up a strong base of support in the West, and we felt the people of Ballina would come out in strength. We had a campaign bus with our logo of a red rose, slogan, and a loudspeaker blaring Simon and Garfunkel's 'Here's to you, Mrs Robinson' – how could we resist? – with the great actor Mick Lally doing a voice-over piece. This was the beginning of a countrywide bus tour that would bring us back to Dublin for the final week.

An enthusiastic crowd came out for me in Ballina, in the rain, and it looked good on television. When we travelled from there to Donegal, the campaign began to get lift off: as we went from town to village to town, surprisingly big crowds were turning out. Momentum was gathering.

We began to see polls that increasingly showed that I was on track to win. Our sense of it, whether right or wrong, was that Brian Lenihan was not doing as well as expected. I was gaining ground on him, quietly, consistently. In one or two regions, some polls had me ahead of him.

Then, high drama. First, on a current affairs television programme, *Questions and Answers*, Brian Lenihan denied making a number of telephone calls to President Hillery on an occasion some years previously to try to influence him (highly inappropriately) in the exercise of one of his powers. But in a taped interview given to a politics student some months earlier, Lenihan had candidly admitted making such calls. Following the *Questions and Answers* programme, the *Irish Times* ran the story, and the following day, the tape was played at a crammed press conference. Lenihan, it seemed, had been caught in a lie. He went on the six o'clock news and, looking into the camera in a direct appeal to the Irish people, denied 'on mature recollection' making the alleged inappropriate call to the president. It was unconvincing and made people squirm. Fine Gael tabled a motion of no confidence in the government, and attention shifted to the junior coalition partner, the Progressive Democrats: Would they bring down the government in light of the revelation? Suddenly, a general election became a real possibility, with the likelihood that some Fianna Fáil TDs would lose their seats. Much pressure was put on Lenihan to resign from cabinet. One of the most loyal and devoted Fianna Fáil party members, a TD for almost thirty years and a minister in nine different administrations, Lenihan refused, and so, in a clinical move to save the Fianna Fáil party, and his own skin, Haughey sacked him.

The upshot: Lenihan, a decent and popular man, immediately received a huge bounce in the polls as presidential candidate, in a quite justified outpouring of sympathy. We became extremely worried. We could feel it all slipping away.

But there was to be a further twist. On a popular Saturday lunchtime radio show, only four days before the election, a cabinet minister,

Padraig Flynn, decided to go for a personal attack on me, and took it too far.

As the weeks had gone by and I had been closing the gap in the polls on Lenihan, the negative campaigning against me had increased. Whether by election or consensus in selecting the agreed candidate, Fianna Fáil had held the presidency since its foundation and would not give it up lightly. And there was an element of Irish society that would never support me because of my track record. At times it got ugly. I was accused of wanting to bring abortion to Ireland, that I would set up an abortionist's clinic in the Áras. They also tried to tag me as a radical socialist, searching through records to find statements I had made that they could use. A well-known priest was heard to describe me as 'a Marxist, lesbian bitch'. Most hurtful were the whispers and rumours circulating that my marriage to Nick was a sham, that we had been secretly divorced for years but were maintaining the pretence in order for me to get elected. Nothing was put in print – these days such rumours would be posted on the internet – as no one could make it stick; Bride dared any journalist who raised such a rumour with her to put something in print so we could clean them out when we sued for defamation.

Padraig Flynn, like me from County Mayo, came on the radio and accused me of having discovered 'a newfound interest in family, being a mother and all that kind of thing . . . but none of us, you know, none of us who knew Mary Robinson very well in previous incarnations ever heard her claiming to be a great wife and mother.'

Michael McDowell of the Progressive Democrats, also on the radio show, came forcefully to my defence, calling Flynn's comments 'disgusting'. Voters, especially women, swung back to me. They were so irate that they decided that, after all, they *were* going to vote, and vote for me. We heard all kinds of stories afterwards about women who traditionally would have asked (or been told by their husbands or fathers) whom to vote for, breaking that mould and giving me their vote. What a humbling experience.

I'd heard the news about the gaffe as I was enjoying a walk down Grafton Street, campaigning through the throngs of shoppers, flanked by well-known Irish personalities – the actors Niall Toibin and Mick

Lally, singer Christy Moore, and other artists and writers – who were coming out strongly in support of me and were getting a terrific response. At the same time I was also being told that Fine Gael had agreed to a transfer deal, in which it encouraged its voters to give me their number-two votes, which Henry and Adrian had a hand in.

I could feel the scales tipping in my favour and was ready to go on the radio the next morning, adopt a presidential, conciliatory tone, and accept Padraig Flynn's apology – such comments can come out in the heat of the moment, and he'd been apologising profusely ever since. But after a blazing row with Bride, she won out. 'You can accept his apology when the voting has closed, Mary. Trust me on this.' It was my turn to get a bounce. (On the day of the first count, when it was clear that I would win, Bride sent Pee Flynn forty red roses with a note saying, 'Thank you, from the women of Ireland.')

In the final days before the election, trying to make the 'socialist' label stick, Fianna Fáil took out full-page newspaper ads asking, 'Is the Left Right for the Park?' (referring to the Phoenix Park in which Áras an Uachtaráin is located) in a final attempt to destabilise my vote – but it wasn't enough.

The counting of votes took two days. From early morning on the first day of counting, I was getting excited messages – from Adrian in Swinford in County Mayo, from Henry in Galway, and from all around the country. A pattern was emerging; unofficially, the tally-men were saying I was running a close enough second to win on the next count.

And so it was. On the evening of the first count, it was clear that when Austin Currie was eliminated, I would get a sufficient proportion of his transfers to overtake Lenihan and win the election. In a dreamy, limbo state, my close campaign team, exhausted souls, began to celebrate.

The following afternoon, I remember arriving at the count centre in the RDS* in Dublin. The atmosphere was electric, the place was mobbed, and I was swept in to a huge crowd of supporters and

---

* Royal Dublin Society, established 1731. Its premises, located in Ballsbridge in Dublin and used as a major venue, are also known as 'the RDS'.

well-wishers, surrounded by my own personal 'minders' (mostly friends of Adrian's from Ballina), and propelled towards the rostrum, where the returning officer stood ready to declare the result.

Eoghan Harris had helped me prepare a rousing acceptance speech.

Today is a day of victory and valediction. Even as I salute my supporters, as Mary Robinson, I must also bid them farewell as president-elect. They are not just partisans, but patriots, too. They know that as president of Ireland I must be a president for all the people, but more than that, I *want* to be a president for all the people. Because I was elected by men and women of all parties and none, by many with great courage who stepped out from the faded flags of the Civil War and voted for a new Ireland, and above all by the women of Ireland, *mná na hÉireann*, who instead of rocking the cradle, rocked the system, and who came out massively to make their mark on the ballot paper and on a new Ireland.

Brian Lenihan and his wife, Ann, were warm and gracious, but clearly furious with Haughey. All Haughey could say to me, out of the corner of a rather bitter mouth, was 'Congratulations. Your car is waiting outside.' I drew a blank for a moment, until I realised that he was telling me I now had an official state car and garda (police) driver.

That night, there was an emotional celebration party for my supporters and campaign workers. I addressed the crowd and thanked them profoundly for all their work, and their belief in me. But I reiterated that other message: 'As I thank all my friends, I must bid you farewell. As president, I will be president for all the people.'

Our three children had been caught up in the excitement, with us for some of the campaigning around the country and present at my acceptance speech in the RDS. Aubrey hung as close to me as he could, holding my hand when possible. It was time to reassure them. We sat at our kitchen table that evening quite late, eating a delicious supper that Nick's brother Peter had rustled up from somewhere. I wanted all three – Tessa, now eighteen, who had just started at Trinity; William, aged sixteen; and Aubrey, aged nine – to understand that even though I had been elected president, they were my first priority. 'Will we live in that big house, Áras an Uachtaráin?', 'How will we get to school?' 'Will it still be like it was now that you are president, Mum?' Despite their concerns I could tell they were proud of me, and Nick kept interjecting with funny remarks, and mimicking some of the characters involved throughout the campaign, so that, in the end, we fell about laughing. Looking at the three of them, I did have an inner pang about how my new role would inevitably impact their lives. On one thing we were all very clear: we would keep them out of the public eye as much as possible.

One of the garda drivers who drove me for the first few days seemed slightly apprehensive when Nick and I got into the car. Before long, though, we became more relaxed with each other. One rainy morning, stuck in traffic, he said, 'I have to tell you, you're causing havoc in the country.' I looked at him, surprised. 'Oh?' 'Yeah . . . One of my friends told me that when he went home to the missus last night, he was tired, you know, and he said to her, "Make me a cup of tea there, love" and she said back to him, "Make your own tea, things have changed around here!"'

We were staggered by the extent of the messages from well-wishers. Sackfuls of post arrived, some letters simply addressed to 'Mary Robinson, Ireland'. Bouquet upon bouquet of flowers arrived, quickly filling our home to capacity. We laid them out on the steps up from the hall to my office, and began sending them on to hospitals and old people's homes.

Then, as the presidential inauguration would not be for another four weeks, Nick and I, exhausted, daunted, took a few days off alone together and went to France to rest and recuperate. As we unwound, we began to imagine what life was going to be like. One stark reality was brought home to us straight away when we discovered to our surprise, that we would be escorted throughout our holiday by an armed security detail of the French police. Strolling back to our hotel one evening from a restaurant nearby, Nick wondered, 'Who are we being protected from?'. Then, laughing, he took my hand and began to pull me along, saying, 'I wonder what they'd do if we made a dash for it?'

During our holiday in France, I had a meeting with the Taoiseach, Charlie Haughey, in Paris. Very high on my list of things to do as president-elect was to create a new position of personal adviser to the president. I needed somebody to guard my back, a trusted adviser. (To my delight and relief Bride Rosney had said she'd be willing to do it.) Another particular item that I wanted to raise with Haughey was an increase in the president's official expenses. I anticipated some hostility and was prepared for a battle over both items. But, surprisingly, Haughey actually suggested an increase in expenses himself – and beyond what I was hoping for. Haughey being Haughey, that was one of his strengths. He believed that people in high office should be able to entertain visitors well. And when I raised the issue of wanting Bride as my personal adviser, he agreed straightaway, no questions asked (a decision he may have regretted later). He was courteous and supportive of me, encouraging me to seek whatever staff changes I felt necessary. 'You're in charge now.' He also reassured me, in no uncertain terms, that it was the Taoiseach's role to defend the president and that he intended to fulfil that role vigorously.

Ours is a relatively young republic; I was Ireland's seventh president since Douglas Hyde entered office in 1938. Writing my inauguration speech was an opportunity to consolidate in a single document my own ideas of the presidency, and my plans for it. It was my solemn declaration to the people of Ireland: I would set out in a formal way a blueprint of what I hoped to achieve during the course of my term. I wanted to capture the idea of an Ireland that was opening up, that was able to elect a woman with a track record like mine, and that at the same time needed to be reminded of its own history. I didn't want to be a president of an Ireland looking only forward and ignoring our rich and wonderful past. Neither did I want to be a president who was smart and liberal and fluent, but distant from the people. I wanted the Irish people to feel that I was representing them and making them proud, but that I was from within, one of them.

I had some help from my old friend, the poet Eavan Boland, and from Professor Richard Kearney of UCD who had written about the concept of the fifth province, with which I opened my inauguration speech. The old Irish term for province is *coicead*, meaning a 'fifth', and yet there are only four geographical provinces on this island. The fifth province is not a geographical place but rather an imaginative space, 'a place within each one of us – that place that is open to the other, that swinging door which allows us to venture out and others to venture in'.

Nick, though, was my primary collaborator in drafting the speech, and it was finished the night before the inauguration by the two of us, sitting up in bed. Right up to the last minute, we were refining it. It was an opportunity to engage imaginatively with language, but mostly, to me, it was an extraordinarily important text because I would try to live by it for the next seven years.

The presidential Rolls-Royce, a 1947 Silver Wraith – first used by President Seán T. O'Kelly but linked in people's minds with his successor, Éamon de Valera – came to collect Nick and me at our home. A large crowd of friends and neighbours had gathered to cheer us as we came out of our front door and got into the car with my aide-de-camp, Colonel Paddy McNally, to whom I had been introduced a few days earlier.

As we drove towards Dublin Castle, where the ceremonies would take place, it was an incredible and humbling experience to see crowds in the street waving, and to see some who looked to be weeping – the moment was just so special. Afterwards so many people said to me that when they saw those images on the television, or heard me taking the oath, or inspecting the guard of honour, they wept.

We were led by the chiefs of the armed forces into St Patrick's Hall to the sound of trumpets. Nick and I walked up to the podium; he found his spot, and I went around the table and stood at the large chair in the centre. It was extraordinary to look out at all the members of the Supreme and High Courts in their wigs and gowns, to see in formal dress the members of government, members of the diplomatic corps, those who represented different aspects of the business community, trade union movement, personal friends such as Seamus

Heaney and Eavan Boland and others I had invited, but most of all to see my father, my brothers, my three children, all smartly dressed. The official party behind me on the platform included the Taoiseach, Charles J. Haughey; and the chief justice – the very friendly face of Tom Finlay – who would administer the oath; the former President Paddy Hillery; and former Taoisigh, including Garret FitzGerald, a tear running down his face during the ceremony.

I was very conscious when I got up to deliver my speech that there had been no great tradition of significant inaugural addresses in Ireland. I had looked at the earlier short and rather formal acceptance speech of President Hillery, and the rather more elaborate one of President Erskine Childers, but I was aware that the speech I was giving was different. I really wanted to deliver it well: it was a promise to the people of Ireland who had trusted me by electing me president, that I would do my very best to fulfil their trust. I had particularly rehearsed the part of it in Irish that talked about my journey around Ireland, as it was important to me that it should be as fluent as I could make it, not being a fluent Irish speaker. I was moved by the warm and enthusiastic response to the speech. And then I took the quill pen that Charlie Haughey had lent me, one that had been used on a previous occasion by President de Valera, and signed the official oath.

Almost immediately afterwards I was escorted outside by the Taoiseach to inspect the guard of honour. I stopped to put on a coat, and to remember, briefly, where I should be standing, and the protocol for inspection, which I had rehearsed in my mind. I was very conscious, as I walked along inspecting the guard, that I was the first woman to be commander-in-chief of the defence forces.

Then I became aware of the cheering crowds that had gathered in the grounds of Dublin Castle. I moved to greet them. It was evident that they, too, had been struck by the image of a woman inspecting the guard of honour. It was then, many people told me afterwards, that they somehow understood that, for the first time, a woman was exercising the power of the presidency.

We were then driven to Áras an Uachtaráin for a lunch, which Nick and I were (technically) hosting for members of the government, council of state, and some close relations and guests. The lunch took place in a combination of the main reception room and dining room. A horseshoe-shaped table had been set up, and I sat at the head of it, with Haughey as Taoiseach on my right and Nick on my left. When the main course was served, Haughey chose that moment to lean in to me and say, 'President, next time I would advise outside caterers.' This was the first indication that Áras an Uachtaráin was not at the time well-geared to host such a function.

There were times during that lunch when I seemed to float above the conversation. I looked out the dining room window at slanting rays of sunshine and thought, *Now I have been elected by the Irish people, I have been trusted by them, and how am I going to fulfil their trust?* I felt an overwhelming sense of responsibility, and at the same time a personal fulfilment that I was being trusted in this way.

Later that afternoon we held a tea party for all the children: my own (their first visit to the Áras) and their various cousins, our nephews and nieces. It was a family event that included the friends whom we had invited to the inauguration. I recall Garret FitzGerald setting off like the Pied Piper as he led many of the children out to the garage to inspect the old Rolls-Royce.

Then it was time to get ready for the formal state reception in Dublin Castle.

We drove into the courtyard in Dublin Castle where a bank of photographers was waiting, got out of the car, and went inside, where we could hear the crowd gathered in the huge reception room: loud voices, laughter, the clinking of glasses. In a quieter, private room the Taoiseach and members of the government had gathered, and some family members: my father, Henry and Adrian and their wives. An understandable lack of enthusiasm on the part of these Fianna Fáil government members was set off by the sheer, wide-eyed giddiness of my family. I enjoyed the moment with my father and brothers, teasing, laughing, and then we went back into the throng, to meet people and circulate before we were called to take our seats for dinner in St Patrick's Hall.

During a gap in conversation, I sat back for a moment to savour these feelings on this extraordinary day in my life, and thought about the final words of the speech I had given earlier.

May God direct me so that my Presidency is one of justice, peace and love. May I have the fortune to preside over an Ireland at a time of exciting transformation when we enter a new Europe where old wounds can be healed, a time when, in the words of Seamus Heaney, 'hope and history rhyme'. May it be a Presidency where I the president can sing to you, citizens of Ireland, the joyous refrain of the fourteenth-century Irish poet as recalled by WB Yeats: 'I am of Ireland . . . come dance with me in Ireland.'

10

The Smell of Fresh Paint

No one moulds us again out of earth and clay,
no one conjures our dust.
No one.

Praised be your name, no one.
For your sake
we shall flower.
Toward
you.

Paul Celan, 'Psalm'
(Translated from the German by Michael Hamburger)

NICK AND I had decided that the family would take up residence above the formal rooms in the main house of Áras an Uachtaráin. As this was not where the Hillerys had lived, some work would be required – fitting a private family kitchen, for example, where we would breakfast – before we moved in properly. We brought with us our trusted housekeeper, Laura Donegan, to help maintain privacy for the family, and she alone, as staff, had access to our living quarters. We hoped to maintain some level of normality for the children, although that wouldn't be easy. William and Aubrey were still attending school and would now be taken there each morning and collected each evening by a garda driver. Tessa, having just started at Trinity, had joined a student flat in town as well as having her room in the Áras.

From the very beginning Nick committed to being as supportive as possible, both of me and of the office, overcoming considerable shyness to do this. He would joke at times in a self-deprecating way about being the first 'first laddie', but of course his emotional support was invaluable to me. To avoid potential conflicts, he had retired from legal work on my election – though, for many years, he had enjoyed chairing the trustee company that held Birr Castle, its historic contents, heritage gardens, and its great telescope, 'the Leviathan of Parsonstown' – and retired, too, from his executive role in the Irish Centre for European Law. He continued his support for the Irish Architectural Archive and, with Edward McParland and others, established the Irish Landmark Trust in 1992, serving as its first chairman.

Nick still had the antennae of a political cartoonist, with a sharp eye for the absurd, the pompous, the insincere, and even a rueful sympathy for the iconoclasts who had me in their sights. He has always had an ability to make me laugh, and during the presidency, those lighter moments were even more precious, and happily quite frequent.

As soon as we settled in he began serious research on a strand of the history of political cartoons that particularly fascinated him. The statesman Edmund Burke had been targeted by leading caricaturists in London for more than thirty years until his death in 1797. Nick had already scoured the great collections in the British Museum, the Library of Congress, Yale, and elsewhere in Britain and the United States, and drew, as well, on a fine personal collection of eighteenth-century caricatures he had been forming since his undergraduate days. Soon the big table in our private quarters was overwhelmed by books and prints and photocopies, and Nick spent many happy hours there when not travelling with me to events in Ireland and abroad. Some years later, in 1997, his work was published by Yale University Press, and it made me enormously proud to stand in the famous Long Room of the library in Trinity with a large attentive audience to hear Nick's book, *Edmund Burke: A Life in Caricature*, being launched and praised. It enjoyed excellent reviews by Conor Cruise O'Brien (a leading authority on Burke) and others, and Nick was invited to give many illustrated talks, at Oxford, Cambridge and New York universities; the Middle Temple in London (where Burke studied law) and elsewhere.

Shortly after taking up residence in the Áras, Nick and I sold our Dublin home in Sandford Road. We needed to pay off debts from the presidential campaign, which were exacerbated by the fact that I had not been earning legal fees during that six-month period. It became important to us to try to find a base where we could go as a family and relax in private. As it happened a house I knew from childhood, Massbrook House, became available through an executor sale. It was a bit beyond our budget, but my father was very keen to have us living close by and chipped in to help us acquire our Mayo home in 1994. From the first moment, I felt an incredible joy to be back in Mayo, able to come down for long weekends and holidays. Our children loved the freedom of the grounds, and we were fortunate to 'inherit' two people, who had worked in Massbrook under its previous owner: John Reilly who had known it intimately from the age of fifteen, and Ann Dyra, a widow from the village of Lahardane, who was delighted to have children around to spoil. Both were soon an essential part of the home and sanctuary that Massbrook became and remains to this day.

Meanwhile, I was starting to get to grips with my new working arrangements. The previous president, Paddy Hillery, and his wife, Maeve, could not have been more supportive, gracious, or friendly. And I was very conscious that my ability to implement my promise and embark on a much more active presidency would be in large part due to the substantial increase in the presidential budget that Haughey had sanctioned.

The president's working office consisted of an outer study, where the president performed official functions such as signing acts of the Oireachtas into law; holding briefings on presidential business with the senior civil servant of the President's Office, the secretary to the president; and meeting with the Taoiseach for regular briefings. An inner study adjoined it and served as my private office, to which no one had access except for Bride, Ann Lane, and Laura (who would come in daily and tidy it for me).

I had been advised by the Taoiseach that I might wish to choose a successor to the long-standing incumbent secretary, who was of retirement age and who, friends had warned me, might be too old-fashioned for the style of presidency I was advocating. Sure enough, on one of my first days in office I was notified that a bill had been sent over from the Oireachtas that I was required to sign into law. (Generally this is a rubber-stamp job, although the president does have the power – a real power – to refuse to sign and to refer the bill to the Supreme Court if he or she has any concerns about the constitutionality of the proposed legislation.)

The secretary came into my outer study with the bill and began insisting that I sign my name in Irish, 'Máire Mhic Róibín'. This seemed unnecessary to me. I had never signed my name in Irish, my signature was 'Mary Robinson', and that was how I had been elected. I told him as much, and still he demurred, citing tradition and insisting that every previous president had done it. Trying to pull the document back off my desk, he said, as if in a scene from the British sitcom *Yes Minister*, 'We'll come back to it later, Uachtarán.' Holding on to the document, I read it over carefully and then signed it Mary Robinson. When he had gone I called Bride and informed her that we should take the appropriate steps to seek a new secretary, less rooted in traditional ways, so that from the beginning the style and approach of the office would be different.

The appointment of my secretary was a key decision, and I took a chance and invited the then-deputy secretary, Peter Ryan, to take on the responsibility. Peter was quite taken aback; somehow he'd presumed he was out, too. He was rather nervous about the whole situation: here was I, come with a fresh broom to the office, and how was change to be effected while keeping within the parameters of the Constitution?

It was one of the best decisions I made as president. Peter Ryan was an excellent secretary to the president: loyal, highly intelligent, hard-working, willing to give advice even if it was not going to be welcomed. He knew the job inside out, and adapted quickly to the more active, inclusive style of presidency I was operating. He never misled me and never (as a civil servant can if he wishes) made things awkward.

Peter's initial relationship with Bride was cool and formal, with much 'Dear Mr Ryan', 'Dear Ms Rosney': she was wary of him, and he of her. Gradually she began to call him Peter, but he still referred to her as Ms Rosney. Then, a month or six weeks into the job, she called into his office in the secretariat seeking advice on a letter. He took the letter, read it, told her what she should and shouldn't do, and then said to her, 'I'm very glad that you didn't show me that letter, Bride, because if you had done, I would have had to take action.' Suddenly he'd called her by her first name and told her that, as a civil servant, there were things that he couldn't and shouldn't do. From then on they formed a warm, collegial friendship.

In those very early days, the atmosphere in the Áras was gloomy. Understandably, the place had a reclusive feel to it, having been out of the public eye for many years, and moved to its own slow rhythm. During meals, even those involving only us as a family, the household staff would come in wearing full uniform and white gloves, one standing behind each chair as we ate. When we found ourselves whispering to each other, inhibited and uncomfortable, I knew that I had to make changes, and quite dramatic ones. I would be doing more entertaining, and of a different type. I needed a more dynamic, professional team. We brought in a highly experienced housekeeper, and a decision was made to change household staff. Some staff were redeployed within the Office of Public Works, their employer – nobody lost his job – and the OPW hired a new team for us, all from the official register of unemployed. But it was poorly handled, and for those redeployed, it became a big union issue, and Haughey – who had prompted me to make whatever changes I wanted – allowed the criticism to mount with no defence or explanations offered, even as I suffered through exaggerated, damaging press reports. I could say nothing in my defence – it would have been inappropriate to do so, he said – and it was infuriating.

During that time, Sister Stanislaus Kennedy, a highly respected activist on the issue of homelessness, and known affectionately as Sister Stan, came up to see me about her latest project. I confided in her about this episode and how painful it was, and she told me, in no uncertain terms, that I was doing exactly the right thing, that I had to be able to do this. This was reassuring, as she was someone of the highest integrity. I rode it out, and it was quickly apparent that the house had been transformed. Making the changes we did enabled us to open up the Áras and to cater for all the different types of groups who would visit. It was a baptism of fire, but worth it.

Near the beginning of my inauguration speech I had said, 'My primary role . . . will be to represent this state. But the state is not the only model of community with which Irish people can and do identify. Beyond our state there is a vast community of Irish emigrants . . .' I knew I wanted to reach out to the more than seventy million people on the globe who claimed Irish descent. 'Typical of a woman,' one wit commented, 'she's elected by three and a half million and she immediately says she wants to represent seventy million.'

I had been concerned over a number of years as a senator and lecturer visiting the United States to see centres in Boston, Philadelphia, and New York having to cope with many Irish who were down on their luck, most of them undocumented immigrants. Even more so in Britain: in Manchester and Birmingham and London, I had seen the poverty, mental illness, and depression of forgotten Irish emigrants. During the presidential campaign, I had gone to London with Dick Spring to draw attention to their plight. Now, as president, I wanted to highlight it.

I knew well the words of Eavan Boland's poem, 'The Emigrant Irish'. 'Like oil lamps we put them out the back/of our houses, of our minds.' When I was growing up in Ballina, it was traditional, coming up to Christmas, for people to light a candle and place it on a window sill at the front of their house. My mother used to make quite a big deal of it. She had a candelabra for five candles that she put in the dining room window. She explained to us, as children, that the lit candle meant that you would give hospitality to a stranger in need.

The idea of the light in the window appealed to me. I had promised, in my acceptance speech, that 'there will always be a light on in Áras an Uachtaráin for our exiles and our emigrants.' For health and safety reasons, of course, we couldn't have a naked flame, so we had a lamp made in the shape of a candle – with no 'Off' button – and put it on the window sill of our family kitchen upstairs, which was under the main portico of the house and could be seen from the public road. I hoped that emigrants would learn of the symbol and know that we at home in Ireland were thinking about them, felt connected to them, and understood if they were lonely or in need of support.

I do remember one stormy night when we had a rare (and brief) electricity blackout in the Áras, and our son Aubrey got up and found a real candle, placed it on the window sill, and kept it lit until the electricity was restored – he was anxious to ensure the continuity of that light.

My thinking about emigrants was simple: we needed to care more about those Irish who had had to leave our shores, and their descendants, and we had to gather in that sense of a wider Irish family.

During my seven years as president, I had the extraordinary privilege of meeting the Irish whenever I travelled abroad. At events arranged by the local Irish embassy or consulate, they came out in incredible numbers. I witnessed this everywhere: in African countries, Argentina, Australia, parts of Europe, the United States, and Canada. I was astonished by the number of times I was welcomed with the words 'We know you have a light in the window,' and by how much that light mattered.

When they were visiting the Áras, Irish descendants who had become leaders in their adopted country – such as Speaker Tip O'Neill, of the US Congress – invariably asked to see the light. We'd go out from the reception room and down some steps to the formal gardens at the front of the house. I would turn around and point it out, in the window, under the portico, between the columns. Then I would turn back and often see misty eyes. Tip O'Neill had tears in his eyes. It brought home to me the powerful link that those with Irish heritage have with Ireland. The Jewish Diaspora, of course, is very special. And perhaps for Polish people there is the same pull back to Poland. For the Irish, I think it comes from the suffering and struggle of Irish history and a spirit of resilience. What putting a light in the window symbolised, and talking about a wider Irish family in my inauguration speech signalled, was the need to light a flame of connection with Ireland.

The emigrant Irish remained a focus throughout my presidency, as did a different level of community. 'Not just the national, not just the global but the local community . . . groups thriving on a new sense of self-confidence and self-empowerment.' As president I would seek to the best of my abilities to promote the growing sense of local participatory democracy I had discovered all over the country when I was out on the campaign trail.

It was a daunting task. In the first days after my inauguration, I felt very isolated. I felt cut off from the interaction I'd had with so many people during the campaign. Still figuring out the limits of my office, I was initially tentative about how exactly I would exercise the more proactive role for which I had canvassed. When, the day after the inauguration, I visited a housing programme for the poor organised by the charity Threshold, I decided to meet people and be photographed but not make a formal speech. But feeling my way into the role, I carved out space and found that there was no reason why I couldn't make speeches at such events. I was determined to fulfil the mandate I had been given in a way that was absolutely correct and within the bounds of the Constitution – I still carried a copy of the Constitution with me, in my handbag, wherever I went – but also to show that a president could do much more.

What also struck me – and I found this extremely difficult in the early days – was the constraint of being first citizen. I had to be first, always: sit in the front row, generally alone; take precedence over such figures as the Taoiseach, ministers, the chief justice. I'd walk through a room and it was as if the waters were parting in front of me. It took some getting used to.

From day one, letters started to arrive from community groups and organisations inviting me to attend the ten-year anniversary of this, or the occasion of that. One day, early on, about three invitations in a row arrived with – as was the standard procedure – the advice from the secretariat clipped onto the side. In each case the staid advice read, 'This event is not of sufficient importance to warrant the presence of the President'. I sat there in my office in the Áras and asked myself, *who is deciding this?* From then on I would decide what was appropriate or of sufficient importance to warrant my presence.

Bride and I were strategic, then, in determining the events I would attend, noting but often overriding official advice. People were savvy, too, and many invitations came in the back door, directly to Bride, rather than through official channels. We worked, sometimes, on the principle that it is easier to seek forgiveness than permission. And as Peter grew into his job, he adapted quickly and became a great ally.

Bride had a map of Ireland on the wall of her office and stuck a pin in every place I visited, to ensure that each county got its fair share of presidential visits. There were funny moments. One of the early visits was to Cavan. For a presidential visit, you have to imagine a motorcade with the official car with the Irish flag waving from one side of the bonnet and the presidential flag – gold harp on a blue background – waving from the other, followed by a second car of protection officers from the special detective unit. Often there would be garda motorcycle outriders with blue lights flashing. For this early visit, then, Cavan County Council, the local authority, repaired at short notice some notorious potholes along the roads where the motorcade would be passing. This in itself was a victory for the local community!

By going to so many local community events, with the flags flying and motorcycle outriders, I was doing something that previous presidents hadn't had the financial capacity or the encouragement to do. Also, consciously, I was linking the presidency with local self-development. If you hold high office, there is power in encouraging, listening, valuing and talking up projects, for by doing it in one place you also communicate with a lot of other people doing similar work elsewhere, who then see it as being more important than they did the day before. It was a significant way to value the spirit of *meitheal* that was changing the local face of parishes and towns and villages in Ireland.

When I was asked, at the end of the presidency, what would be my abiding memory, my answer was 'the smell of fresh paint.' There were so many places that were either new or had been given their 'makeover' for the visit of the president.

It is not uncommon for non-executive presidents to require government approval in order to leave the jurisdiction, and the Irish

Constitution, in Article 12.9, directs that 'The President shall not leave the State during his term of office save with the consent of the Government.'

My first trip outside the state was in January 1991, to attend the funeral of the much-loved King Olaf of Norway. It was a very moving ceremony. Inevitably there was quite a bit of curiosity about me as this newly elected woman president of Ireland, and I received a number of requests from the international press for interviews. Attending with me, the minister for foreign affairs, Gerry Collins, advised strongly that I should not give any interviews. I had not intended to, as I felt it would be inappropriate given the occasion, but I noted the advice, and the minister's body language indicating that it would never be appropriate for the president to talk to the press, and tucked it away in my mind to think about.

The government's consent for that trip went through so smoothly that I barely thought about it. But when I was invited, shortly afterwards, to give the prestigious annual Dimbleby Lecture in London, founded in 1972 in memory of BBC broadcaster Richard Dimbleby, government consent was refused, on the grounds that it was 'not appropriate'.

Not long after that a request came from the Taoiseach's office: he wanted to meet me. A worried Peter informed me that he, Peter, had been tipped off by his counterpart in the Taoiseach's office, Frank Murray, that Haughey would be bringing legal advice with him. This aroused my lawyerly curiosity, and a certain sense of pressure. Haughey duly arrived, and was ushered into my outer study, the two civil servants, Frank and Peter, remaining outside. Tea was brought in and served, and then the door closed. Haughey took out of his pocket a legal opinion of senior counsel obtained by the government and read it aloud to me. It gave a staid, conservative interpretation of the provisions of the Constitution to the effect that the president could not give press interviews or speak independently of the government – it would be considered 'addressing the nation' – without approval of the government. A muzzle, essentially.

NEXT IS THE NEWSPAPER
OPINION POLL CONTROL UNIT.

The Taoiseach, Charles J Haughey – as seen by Martyn Turner – rushes, with muzzle in hand, to oversee the President's public utterances.

I reminded him that I was a constitutional lawyer, and then proceeded to refute the argument, offering alternative interpretations of the wording. This was my field, and I was on top of the issue because I had lived and grappled with every word of those articles of the Constitution since the moment I had been elected; that was my binding, guiding source. I could answer every point and Haughey could see that my arguments were more persuasive than the legal opinion that had been furnished to him. Eventually, with a flourish, he threw the opinion to the floor and said ruefully, 'Ah, you get the advice you pay for!'

Later, after the Taoiseach had returned to his office, Peter asked me to meet him in the outer study, where I found him grinning. He told me that according to Frank Murray, Haughey, far from pleased, had conceded the legal opinion. The president could speak at public meetings and communicate directly with the press.

After that there weren't many difficulties between Haughey and me, except for the issue of the Dalai Lama.

His Holiness, the Dalai Lama, was coming to Ireland at the invitation of Mr Justice Brian Walsh, chair of Dublin's Chester Beatty Library, to open a Tibetan collection in the library. We had not met before but I had long admired him and was proud that he had personally written to thank me for forcing a debate in the Senate in 1989 on the Chinese violation of human rights in Tibet. The Irish government couldn't very well refuse the Dalai Lama entry into the country, but Haughey had made it known that no one from the government would be meeting with him. He wanted to avoid Chinese censure.

Advice was conveyed from the Taoiseach's office, through Peter, that the Dalai Lama should not be received in the Áras and that I should not meet him. Meanwhile Bride was directly contacted by the Chester Beatty Library and Brian Walsh with an invitation for me to come along to the event, meet with His Holiness, and go around the collection with him.

I agreed not to invite the Dalai Lama to the Áras, but decided to accept the invitation to the collection opening. I called Peter in and told him I was going to meet His Holiness, although I appreciated it was awkward for Peter, who was being put under enormous government pressure. He gave me the advice he was required to give me then communicated my position to the Department of the Taoiseach. We quickly received a written reply from Haughey urging me not to go. It was carefully crafted, not prohibiting me from going, but couched in the language of there being 'concerns'. On the day of the exhibition, 22 March 1991, I received two further letters from Haughey in that vein.

The stakes were high. We were told that a meeting could affect our relations with China; the Chinese would consider this a hostile act. As far as I was concerned, I was not receiving the Dalai Lama in the official residence of the Irish head of state, and that should be enough to fend off the Chinese 'threat'. I was going to the opening of a collection by the Dalai Lama, and I was perfectly entitled to do that. So we were at an impasse.

It is hard to convey just how stressful the day of the event was. As a human rights advocate, I was keen to meet the Dalai Lama. I was of the view that as he had come to Ireland and I had been invited by a former Supreme Court judge to meet him, it would be wrong for me to refuse. It would be deeply wrong.

However, as the pressure mounted, I felt increasingly sick to my stomach. It wasn't helped by the fact that Peter was getting whiter and whiter and more and more anxious. He warned me that there might be a background to this I was not being told about, that a trap might be sprung. There might be some trade contract that Ireland was about to sign with China involving Irish jobs that I might unknowingly put at risk. Then, less than an hour before I was to leave for the event (which was still up in the air), I was sitting in my inner study – wondering, *Am I going, am I not?* – when Peter, who almost never came into the inner study, knocked on the door and burst in with a broad smile to say, 'It's all over. The Taoiseach's gone to his island!' Haughey had left Dublin to go down to his private island, Inis Mhicileain, for the weekend. He'd thrown in the towel.

The cars came around almost immediately. Even in the car, my throat was dry, my stomach was feeling awful. As we drove up Shrewsbury Road, my driver pointed out two ministerial cars idling by the side of the road, one containing Bertie Ahern, the other Mary O'Rourke, seemingly willing to defy Haughey but waiting first to see if I would go in.

When we arrived, the Dalai Lama was already there. I was brought to a private room so we could have a moment together. He put out his hand and laid it on my lower arm and he said, 'I know, I know.' It was unbelievable. He said something funny, and as we laughed, I relaxed. Then someone came to take us into the collection, 'The Noble Path'. It was a great moment, to be there with the Dalai Lama and see his joy at viewing the masterpieces of his people; his excitement was childlike. He was so straightforward and simple and profound.

My meeting with the Dalai Lama began a very special relationship. For the rest of his time in Europe, he used the photograph of the two of us in his literature. I was the only European head of state to meet with him on that particular trip. I had gifted him with what he needed, recognition at that level.

There was no follow-up from the Chinese.

Despite the refusal of consent to travel to England for the Dimbleby Lecture, invitations to the United Kingdom continued to land on my desk. I was invited to accept an honorary degree from Cambridge University, I was invited to join with President Václav Havel of Czechoslovakia to adjudicate on a competition to design the logo of the new Bank for Reconstruction of Eastern Europe, which required a meeting in London, and I received many requests from the Irish community in Britain, from groups I had visited during the presidential campaign. So I submitted these invitations to the government for permission to travel outside the jurisdiction, believing that they could not keep refusing. This threw the government into consternation. It is difficult to believe now, but no previous Irish president in the history of the state had travelled to 'mainland' Britain for official business. There was a problem, for example, with how to describe me. Because of Northern Ireland, the British would not agree to call me 'President of Ireland'.

Reluctantly, the Irish Government consented to my leaving the jurisdiction in April 1991 to accept Cambridge's honorary degree and to fulfil the engagement with President Havel. I was heavily dissuaded from meeting with the Irish community on that particular occasion, and not wanting to rock the boat too much at that stage, I agreed not to.

We had a wonderful outing in Cambridge: my father, brothers Henry and Adrian, and daughter, Tessa, joined us. We met with the Duke of Edinburgh, Prince Philip, who was chancellor of the university, and it did not seem so outlandishly impossible after all that an Irish president would meet with the British monarch. Sure enough – and notwithstanding the conflict that was still going on in Northern Ireland – two years later I received an invitation from Queen Elizabeth to meet at Buckingham Palace. This was neither a 'state visit' nor an 'official visit'; it was a sui generis, carefully orchestrated meeting at head-of-state level between two countries with a troubled past seeking, in a symbolic way to begin reconciling. The Irish Government very much welcomed this invitation as an important breakthrough, and there was no question of withholding consent to my travelling.

An editorial in the *Irish Times* (22 May 1993) following the announcement of the visit captured this sense of its significance:

> That symbols count as a potent force in politics and international relations is a truth no student of Anglo-Irish affairs can ignore. It is therefore difficult to exaggerate the significance of yesterday's Government announcement that the President, Mrs Robinson, is to accept Queen Elizabeth's invitation to pay a courtesy call on her in Buckingham Palace next week . . .
>
> To symbolise is to represent. Two more appropriate heads of state to undertake this task can hardly be imagined than Mrs Robinson and Queen Elizabeth . . .
>
> This meeting . . . symbolises a maturing process in the British-Irish relationship, a point at which, it is very much to be hoped, disagreements can be resolved within what is a much more amicable framework. Such turning points are often reached when the problems they represent are more nearly capable of solution. There are many objective factors allowing for a qualified hope that this might be the case between Britain and Ireland and within Ireland North and South. It is the subjective factors which may present the major barriers to a settlement. This meeting could do a lot to establish an atmosphere of greater trust in which it would be easier for them to dissolve.

I was conscious of the symbolic importance of this visit, that it was a moment of history. I wanted to look well – one of the few times I really cared about my outfit; everything had to be just so – and be relaxed, and not nervous, as I represented the Irish people.

Our son William was studying at Strathclyde in Scotland at the time, and had travelled down to join Nick and me, Peter and Bride, and officials from the Department of Foreign Affairs. We drove in to the front of Buckingham Palace and then to a side door. Nick and I were immediately taken upstairs to a drawing room. The Queen was there and greeted us warmly, and after tea had been poured we were left on our own with her. There were three teacups, a larger one for Nick than for us two women. I felt confident, pleased that the meeting was taking place, happy to be there, and I conveyed that. I

"Admission is £8 each, Madam"

Scratch (Aongus Collins) marries two stories to create a classic cartoon: Queen Elizabeth's invitation, and the announcement that members of the public would shortly be able to visit Buckingham Palace.

thought the Queen was slightly apprehensive, perhaps of some issues that I might raise. When I said to her that I hoped that at some point in the future she would come to visit Ireland, her face lit up and she responded warmly. It was quite obvious that was something she was interested in doing.

We knew in advance that, in a break with normal protocol, the Queen had agreed that there would be a photo opportunity for the large contingent of press outside. She was acknowledging that this meeting was a significant one for both countries and needed to be recorded. We came out of the very fine drawing room and into a lift that the Queen worked herself. As we exited the lift, there, waiting in a row to be introduced, were William, Bride, and Peter. The Queen and I then walked to a doorway where we could hear the press outside, an enormous bank of photographers. William instinctively stood back. Nick was photographed with us initially and then he, too, stepped back so it was just the two of us. That was the photograph that would be on the front page of all the newspapers the next morning. Then, seamlessly, our Rolls-Royce drove up, and we took our leave, thanking the Queen for her hospitality. Nick got in on the far side, and off we drove. (Poor William, unused to the ways of the motorcade, found himself stranded with the Queen and not sure what to do. So he shook her hand, mumbled something like, 'Very nice to meet you', and ran to find space in one of the cars at the rear!)

The Irish embassy in London hosted a huge event that evening. People there of every background told me that my meeting with the Queen would have a terrific impact on their working lives. I had feared some might be cynical about the meeting but it was exactly the opposite. A taboo had been broken: the president of Ireland and the Queen of England had stood side by side, two women, heads held high. Somehow, for the Irish psyche, that was a very positive symbol.★

I paid an 'official visit' to Britain, in 1996. The distinction between 'state visit' and 'official visit' is that I was invited by the prime minister, John Major, as the guest of the government, rather than by the Queen. Relations between the countries were steadily improving, but with much still to be done to secure peace in Northern Ireland, a state visit would have been premature.

This official visit was an important occasion and a happy one: a breakthrough, with an official lunch in Downing Street and, the next day, lunch with the Queen and Prince Edward in Buckingham Palace. This was preceded by a highly symbolic event: my inspecting a guard of honour in the courtyard of Buckingham Palace. The regiment of the Irish Guard was lined out in full dress uniform, standing to attention in their distinctive bearskins, and I, president of Ireland, on behalf of a proud nation, inspected them, to the rousing beat of the Irish National Anthem, 'Amhrán na bhFiann'. It sent shivers down my spine.

★ Years later, in 2011, Queen Elizabeth came on her historic state visit to Ireland. My distinguished successor Mary McAleese, welcomed her, and the two women stood side by side, in like manner.

11

Hope and History

History says, Don't hope
On this side of the grave.
But then, once in a lifetime
The longed-for tidal wave
Of justice can rise up,
And hope and history rhyme.

Seamus Heaney, 'The Cure At Troy'

ON 10 February 1992, Charles Haughey resigned as Taoiseach under a cloud of allegations about a phone tapping scandal in the 1980s that had come to a dramatic head. As is the protocol upon resignation, he came to Áras an Uachtaráin to relinquish his seal of office. We took our usual seats in the outer study, as we had done on the occasions when, as Taoiseach, he came to brief me, but this time our conversation was less formal and more personal. Later that evening, Albert Reynolds paid his official call to be presented with the same seal of office.

When the ceremonies were over, I ordered a bottle of champagne from the kitchen and invited Bride and Peter to join me. We discussed the events leading up to that day, and some of the trickier times we'd had with Haughey. There were plenty of those, but at the same time I had to acknowledge that the increase in the presidential budget that he had agreed to so readily had helped me be much more proactive as president, and I was grateful. After Haughey, and during my term as president, a further three Taoisigh would serve: Albert Reynolds, John Bruton, and Bertie Ahern, each with his own distinctive style and political agenda, but none would prove to be adversarial in the way Charlie Haughey had been, or would try to control what I was doing.

One of the significant constitutional powers of the president was, with the consent of the government, to address both houses of the Oireachtas. I did this on two occasions, on 8 July 1992, on the theme of 'Irish Identity in Europe', and on 2 February 1995, on 'Cherishing the Diaspora'. Until then the word *diaspora* was not used to refer to the vast number of Irish emigrants and their descendants. Some people I consulted felt it would not be understood, so, needing reassurance, I sounded out my father. 'Definitely use the word "*diaspora*" Mary,' he said. 'The Irish love new words.'

He was right about that, but my speech was not well received by the deputies and senators I addressed. I could feel their resistance, and the applause was tepid. This was a sensitive area because of my implied criticism that successive Irish Governments had failed to support either emigrants or the centres in Britain and elsewhere that looked after them. The high level of emigration from Ireland, driven by lack of job opportunities at home, was itself a reproach. In retrospect, I can also see that my tone was preachy and lacked humour or lighter moments. I felt deflated as I left the Oireachtas, but then messages – e-mails, letters, phone calls – began to pour in from around the world. The address had been carried live on Irish television, and had been picked up by media internationally. I received handwritten notes from nuns and priests in remote parts of China, Africa, Latin America, thanking me. Senator Ted Kennedy sent a note that he had placed the speech on the official record of Congress.

Although as president I would have to be circumspect, and not become politically involved, I placed great priority on supporting those elements of the communities in Northern Ireland that were rejecting violence and seeking out a peaceful resolution to the conflict. For that reason, I had visited Northern Ireland as part of my presidential campaign. Because of the work I had done as a human rights activist, questioning the validity of the Special Criminal Court and opposing overreach in the Offences Against the State Act, I felt I had some credit with the nationalist and republican communities; and my stance on the Anglo-Irish Agreement in 1985 gave me an entrée with the loyalist and unionist communities, at many different levels.

I had decided, very deliberately, to speak of my affection for Northern Ireland in my inauguration speech and to call it 'Northern Ireland': in 1990, many people from the Republic referred to 'the North', whereas to call it 'Northern Ireland' recognised it as a separate and distinct jurisdiction. This was before any peace talks, before the Peace Process, during a period of often violent conflict, where implementation of the Anglo-Irish Agreement was encountering manifold difficulties.

I said, 'As the elected choice of the people of this part of our island I want to extend the hand of friendship and of love to both communities in the other part.'

This was strong language – using the word 'love' – and I knew it was unusual for someone making a public speech. But I wanted to extend this hand of friendship 'with no hidden agenda, no strings attached'. I was confident I could reach out because of my track record with both communities.

From early on in my term as president, I invited groups from Northern Ireland to visit Áras an Uachtaráin, especially those with a cross-community or North-South emphasis. Women's groups were strong on this, and I remember particularly a visit of women from a mix of both Catholic and Protestant housing estates, who came by train, many of them visiting Dublin for the first time. They had dolled themselves up to the nines and came into the Áras in a burst of excited, northern voices. We sat around the large table in the formal dining room, and as there weren't enough chairs, some sat on the floor. I was struck by their sharp wit and sense of humour, their constant cracking of jokes, these women of resilience who had coped for so long with the violence, the discrimination, the high unemployment levels in their communities.

As it happened, my first visit to Northern Ireland, which was for the installation of Cathal Daly as Archbishop of Armagh, took place very shortly after my inauguration, in December 1990. That was organised by the government as official state business, and the crossing of the border, and the visit, passed without incident. Unused to it all, I endured that quite isolated feeling of sitting at the very front of a packed cathedral, in a row on my own, but was greatly cheered by the warmth from both priests and citizens as I came out of the cathedral, a goodwill that I felt I could build on.

My next visit to Northern Ireland, in February 1992, was organised from the president's office rather than by the government. We decided that I would accept an invitation from a coalition of women's organisations from the most disadvantaged areas, and from both traditions – the Falls Women's Centre, the Shankill Women's Centre, the Craigie Women's Centre, the Greenway Women's Centre – to take place at the Equality Commission offices in Belfast. The meeting was organised with the help of an old friend and ally, Inez McCormack, a leading trade unionist and champion of equality issues who played a constructive role in combating discrimination against Catholics in Northern Ireland. Inez and I had sat on many panels together over the years; she was popular with women activists both for her natural mirth and for her steeliness in dealing with adversaries.

Inez and Bride constructed a visit that was focused on inclusivity: the coalition of women's groups acted as the hosts of the event, and they, turning normal practice on its head, invited the 'great and good' of Belfast – officials, dignitaries, elected representatives, and so on – to meet with me at this reception; the outsiders inviting the insiders.

Some journalists didn't understand the point of this and wondered, because it wasn't really political, why I was 'wasting my time on something so petty'; others, notably the late Mary Holland, the *Irish Times* Northern Ireland correspondent at the time, did get it. Most important, the women's groups hosting the event understood the tone of what I was doing: I was there to respect, to affirm, to recognise their struggles, and also to urge them to move on a bit and to shrug off as much as they could any vestiges of victimhood.

The visit had posed the question of whether I needed government permission to travel. Should I concede that I was leaving the state by going to Northern Ireland? We decided not to make an issue of it, and so, as a courtesy, we 'informed' the government that I would be travelling to Belfast for this event. And the government of the day was happy to collude in that fudge; no Fianna Fáil government was going to take an official line that by travelling to Northern Ireland I was leaving the jurisdiction.

It seems difficult to credit now, but that visit was the first working visit by an Irish president to any part of Northern Ireland. Other trips to Northern Ireland followed and would total eighteen. The most significant, certainly the most controversial, was my visit to West Belfast in June 1993.

It is virtually impossible to describe the intensity of that occasion without understanding the context. There was no cease-fire and no peace process at that stage, and the violence was escalating. The situation was at an impasse, framed in terms of 'win or lose'. The British Government was determined that it would 'win the war' but the IRA would never concede 'losing' it.

The British and Irish Governments and the Northern Irish establishment took the trenchant view that to say anything other than 'defeat the gun' was to support the gun. There was no room to bring any other perspective to the situation, such as urging that a human rights framework could facilitate a dialogue without blame being attributed. That came later, that breaking of the mould, through the work of human rights-oriented organisations such as the Committee on the Administration of Justice, based in Belfast, with the help of international and US-based groups, notably the Lawyers Committee for Human Rights (later Human Rights First), led by Mike Posner. They set a different agenda for discussions, one that allowed a scrutiny of government and of establishment obligations and practices in light of human rights standards.

But in 1993 the dynamic was that those using and supporting violence had to be isolated. There was no question of the state taking responsibility for, or taking part in, any dialogue or solution.

My practice of inviting groups and community representatives from Northern Ireland to visit Áras an Uachtaráin resulted, in turn, in invitations to travel to visit them, including an invitation to visit Republican West Belfast. Bride contacted Inez McCormack to see if she could 'get me in' to West Belfast. Inez's immediate reaction was yes, but I'd have to meet the local elected representatives there, one of whom was Gerry Adams, leader of Sinn Féin, the 'political wing' of the IRA. If I tried to go without meeting Adams, it would be wrong; I would be disrespecting the community.

I knew this was going to be difficult. Nobody was going into West Belfast and nobody was meeting with Gerry Adams; he was still, at that time, subject to the broadcast ban, meaning it was illegal to broadcast his voice on radio or television. This community in West Belfast felt completely isolated: in its view, it was not a part of Britain. It flew the Irish flag, spoke the Irish language, asserted its Irish identity, but because of the taint of violence, of the gun, nobody in the Republic of Ireland was reaching out. No money was going in to West Belfast from anywhere; it was a no-go area for the RUC, the then-Northern Irish police force, and was effectively self-policed by the IRA. If the RUC did go in to the area, there would be an incident. Yet it had a vibrant community; it was full of good people working hard to counter the lack of facilities and resources, and the discrimination they suffered. That was what I wanted to honour.

Inez and I decided to construct an event in that vein: I would go to listen and to see, to learn and to value the resilience and strength of community development in West Belfast. It was agreed that I would meet Gerry Adams, but – to avoid the visit being hijacked by any triumphalism on the part of Sinn Féin – privately, without a photographic record.

Albert Reynolds was Taoiseach of a Fianna Fáil–Labour coalition government of which Dick Spring was Tánaiste and minister for foreign affairs. When I informed the government, and these two office-holders in particular, that I would be accepting an invitation to travel to West Belfast, they were not a bit pleased. And the more hostile of the two was Spring. Reynolds was a canny Taoiseach who had a good instinct for peace talks and what would become the peace process. Dick Spring's Labour had performed very well in the 1992 general election (thanks, in part, to the successful presidential campaign in 1990) and had entered into government. There was some residual tension between my office and the Labour Party, I think, because, to a fault, I was so adamant not to show any favouritism. In any case, while Reynolds made it clear that he was not going to stop me, the initial advice I received from the Department of Foreign Affairs was strongly discouraging.

My office was then informed that the British had insisted to the Irish government that the proposed visit be cancelled, that I should not under any circumstances travel to West Belfast and meet with Gerry Adams. I had to think very carefully about this. I wanted to avoid any damage to the office of president that would be caused by my straying into the political sphere, but the purpose of my proposed visit was not to meet with Gerry Adams; it was to meet with and recognise the vibrant community in West Belfast. The reality was that I could not do so without meeting its leaders. The British could argue that I should make this visit while avoiding Sinn Féin, that I could meet with the more moderate SDLP elected representative, Joe Hendron, and avoid the controversy of Gerry Adams's presence. But I knew full well that while Hendron was an elected representative, Adams was the significant leader, and snubbing him would effectively be a snub to the community, would support the politics of exclusion, and would render the visit worthless. What I was trying to achieve was to bring this community a tiny bit in from the cold.

We decided to go ahead with the visit as planned. We were conscious – and Inez McCormack, organising at the Belfast end, was keenly aware – that for this trip to succeed, we would have to have all our ducks in a row. As well as British opposition and Irish Government discouragement, unionists and loyalists in Northern Ireland, and all those who benefitted from maintenance of the status quo, would be vociferously opposed to my entering West Belfast – and it didn't matter that I had criticised the Anglo-Irish Agreement as a senator or that I had so recently been received in Buckingham Palace.

On 18 June 1993 we drove up to the border and were handed over to our Northern Ireland security detail, the RUC. Our first stop was the Down County Museum. Then, as we drove towards the Rupert Stanley College in Whiterock Road, where the West Belfast event was to take place, the security detail literally peeled away from us. Still, whether or not they had been ordered to withdraw, three RUC special branch officers stayed to protect me. I had been warned of the potential personal threat against Nick and me from rogue elements. I believe the officer in charge, whose name was Henry, took an old-fashioned notion that I was on his patch and, however much he might disapprove of my visit – and, indeed, appreciating the potential danger that he and his men might face in this part of Belfast – he was not going to abandon me.

We arrived at the grounds of the college, which was lined with excited schoolchildren waving tricolours. Inez was very anxious lest this might be seen as provocative, but it was really an innocent, exuberant display. I will never forget the palpable sense of excitement when I went into that community hall. It was one of those completely emotional occasions. Everyone was excited and knew that something had happened, some taboo had been broken. The St Agnes Choral Society was singing, the pipes were being played, and there was Irish dancing, a community expression of vibrancy and culture. I met with teachers and trade unionists, musicians and schoolchildren. Then I went into a separate room and, out of view of the cameras, privately shook hands with the local politicians. Gerry Adams welcomed me formally in Irish and thanked me for coming. And that was it. Then I was on to Joe Hendron and chatting with him, and then on to the local councillors. In fairness to Sinn Féin, they kept scrupulously to the deal and didn't hijack the meeting in any way.

Later, Inez described the scene to me. She was standing towards the back, watching me negotiate my way in and around this room full of people. A special branch officer was watching me, and a few steps behind, the head of the IRA security was watching him, and she was watching them both, and watching for body language. She described how I was chatting to one very tough Republican woman, smiling because it was all going beautifully, when suddenly this woman's body went stiff with horror. I had just asked her if I could deliver all the flowers I was receiving to a local old people's home. She beat her way over to Inez and started explaining to her in a panic how she had cleared the roads of IRA and of British Army for this visit, but there was no way she could get me safely to the old people's home to leave the flowers there. 'I'll get the flowers to any fucking place she wants them,' she said, 'as long as she's not going with them!'

I stayed overnight in Belfast because I was doing a cross-community event there the following morning, before giving a press conference at the airport and then flying home. I woke up in my hotel room to find the newspapers full of the event, mostly condemning it. As was my routine, I washed my hair and waited for the hairdresser we had booked to arrive – and waited – until at a certain stage we realised that there was to be no hairdresser. Disapproving that I had gone into West Belfast, she had refused to come. I had no curlers, but dried my hair as best I could. Outside the event there was a picket with protesters, including the activist Bernadette McAliskey. One of the protestors was heard to complain, 'She could at least have got her hair done!'

At the airport in Belfast, I was to hold a big international press conference – there was keen interest from the American media, who gave the event wide coverage in the States. I got word from Henry, the RUC officer, that I should arrive at the airport early as he had arranged a hairdresser for me. I thought that was really professional of him.

The press conference was crowded and difficult. I had to take a careful line. I felt conscientious about never wanting to let the office of president become politicised. But I felt strongly that the local community in West Belfast also deserved the oxygen of recognition that I had been able to give to other communities; it deserved to be valued for the work it was doing on the ground, and not just be typecast as being from Republican, violent West Belfast. And it worked, I believe. It changed attitudes. It brought those communities out from the beleaguered situation they had found themselves in. I didn't regret it for a moment.

I arrived in Dublin to find a number of government ministers at the airport to welcome me back, including Bertie Ahern, which I appreciated because of the extent of the criticism that had been voiced in parts of the media.

Should I have gone? A poll asking whether or not I should have shaken Gerry Adams's hand was published in a Sunday newspaper less than a week later. 'Yes', said 75 per cent of those asked, and my personal popularity rate was recorded as a humbling 93 per cent.

At a time when the voices of Sinn Fein leaders were banned and could not be used for radio and television broadcasts, yet actors could do voice-overs, cartoonist Martyn Turner proposes a solution for my encounter with Gerry Adams in 1993.

There were times when the conflict in Northern Ireland took on a very personal note. One of the guests I had invited to my inauguration was Gordon Wilson, whose daughter Marie had been killed, at his side, in a notorious IRA bombing in Enniskillen on Remembrance Day 1986. Gordon had become a moral voice for reconciliation and healing. He was also a wise friend when another terrible IRA bombing occurred, in March 1993, in Warrington, England, killing two young boys, Johnathan Ball and Tim Parry. At my request Gordon went to Warrington, spoke with the parents of both boys and came to see me in the Áras to say that I would be welcome to attend the memorial service in St Elphin's Parish Church, Warrington, to represent 'the true spirit of Ireland'. I did so, and later returned to Warrington to launch, together with Prince

Charles, the Warrington Project, building links between the various Irish and British communities affected by the Northern Ireland conflict.

Later that year, in December 1993, Taoiseach Albert Reynolds and British prime minister John Major made a historic joint declaration, which became known as the Downing Street Declaration. It followed from secret talks between Gerry Adams of Sinn Féin and John Hume of the SDLP, quietly and 'unofficially' supported by officials from both the Irish and British Governments, and was built on the groundwork of the Hume/Adams Initiative. The declaration signalled the beginning of peace in Northern Ireland. It recognised the right of the people of Ireland, North and South, to freely determine their future relationship, but with assurances to the unionist population that a united Ireland could come about only with the consent of the majority of the people in Northern Ireland.

The declaration went far enough to slowly bring in from the cold those espousing violence, and less than a year later, the IRA and loyalist paramilitary groups announced a cease-fire, thereby manifesting a commitment to enter into talks – the peace process – which culminated in the 1998 Good Friday Agreement.

An issue that mattered hugely to me, of course, was human rights. In my inauguration speech (though, at that stage I still had not worked out how I would do it), I had said, 'Looking outwards from Ireland, I would like on your behalf to contribute to the international protection and promotion of human rights. One of our greatest national resources has always been, and still is, our ability to serve as a moral and political conscience in world affairs. We have a long history of providing spiritual, cultural, and social assistance to other countries in need – most notably in Latin America, Africa and other Third World countries. And we can continue to promote these values by taking principled and independent stands on issues of international importance.'

Even as I was drafting this paragraph I was thinking of the limits of my role as president, yet I really wanted to do something significant in this area. That was my track record – human rights lawyer and campaigner – and being voted into office by the people had created the mandate to pursue this on their behalf, and fuelled an expectation that I would.

In the autumn of 1992 Irish aid agencies – Concern, Trócaire, GOAL – contacted me, through Bride, about the drought and famine in Somalia and how they were desperately trying to get food into parts of the country. Their request for my help to publicise the crisis, and their efforts, sowed the seed of an idea of my actually travelling there, as president of Ireland, to bear witness. It became not a question of would I go, but rather how to make it happen and obtain government consent to travel. In the event, I invited the aid agencies to come to the Áras to brief me at a press conference. The arrangement, orchestrated by Bride, was that, at a particular point in the press conference, Father Aengus Finucane, the head of Concern, would ask me – in this very public setting in front of the press – if I would consider going myself to Somalia, and perhaps reporting to the United Nations afterwards.

It nearly all fell apart. Charlie Bird was covering the press conference for RTÉ television news, and he had decided, before the end, that he had what he needed and was going. Bride had to sit him back down again and suggest to him that he wait another minute or two. So the question was asked and I replied, 'Well, this would be a matter for the government, but of course on a personal level I would be delighted to go.'

In fairness, after some to-ing and fro-ing between Peter Ryan and the Taoiseach Albert Reynolds's office, the government did consent to the trip, notwithstanding security concerns.

The minister for foreign affairs, David Andrews, who had a long record of caring about issues of development, accompanied me, as did Nick and Bride. A large contingent of Irish press also travelled. We flew to Nairobi, and from there, on 3 October 1992, took a small plane to Baidoa, in Somalia. We arrived on a dusty landing strip that was really a field and, in the sweltering heat, were greeted by the elders, the Irish aid agency people there, and the local officials. Young Somali men sat in the back of trucks and jeeps armed with rifles and machine guns. They had been hired to provide security.

At the feeding stations run by the aid agencies, we saw rows and rows of people, men and women and emaciated children. I knelt down beside one woman and asked her to show me her baby. He was tiny and ill-looking with sores on his scalp and flies crawling over his face and eyes. She offered him to me, and I picked him up— he was light as a feather—and tried to murmur some comfort to him. Behind me, I heard a photographer call out, 'Look this way, President' and I turned and posed holding the child, hearing the clicking of camera shutters. But then I suddenly felt ashamed to be playing to the cameras. I knew the photographers had a job to do, that this was a difficult assignment for them, too, but I made a mental decision not to pay any attention to the cameras, but to focus on the people, giving them as much support and visibility as I could.

It was extremely difficult to cope with the misery we were witnessing: children all skin and bones but with bellies swollen from malnutrition and acute hunger; their parents powerless to help them. They, too, were dying, and others were badly wounded because of the fighting, the underlying conflict. And the numbers: row upon row upon row of people waiting listlessly for food and medicine. The sheer scale of the crisis was sobering.

It was a gruelling visit. With the help of the UN special representative, Ambassador Mohamed Sahnoun from Algeria, I was able to meet two of the warlords, Ali Mahdi Muhammad and General Mohamed Farrah Aidid. I urged both to end the violence and allow food to the feeding stations. I really appreciated David Andrews's quiet and skilful support in these talks and throughout. We visited a small field hospital in Mogadishu where we were led straight into an operating theatre where limbs were being amputated without the benefit of anaesthetic. On the last day, we travelled to North Kenya to see the terrible living conditions of Somalis who had fled there.

When we got back to Nairobi, for an international press conference, we were drained and emotional. There was huge media coverage. I had prepared what I was going to say, but as soon as I got up to speak a wave of emotion hit me and my voice broke. I was furious with myself; I was supposed to be telling the world, calmly, coherently, professionally, about the injustices I had seen, but here I was, struggling to steady my voice, to hold back the tears. Beside me, David Andrews was banging the table with the flat of his hand to control his own emotions. Nick, in the front row, was stressed that he was not up on the platform beside me, hating to see me so upset without being able to comfort me. I just about managed to calm myself and speak.

'It has been a very difficult three days. Very very difficult. I found that when I was there in Baidoa and in Afgoi and in Mogadishu and this morning in Mandera I had no real difficulty in remaining calm and in not letting my emotions show. And I find that I cannot be entirely calm speaking to you, because I have such a sense of what the world must take responsibility for. . . .

'I have an inner sense of justice and it has been offended by what I've seen in the last three days – deeply offended.'

Afterwards, when Nick and I got back to our hotel room, I was disappointed with myself. In all my professional career I had never lost control of my emotions like that. Nick was upset that he had not been able to steady me. 'You did okay, Mary,' he said. 'People will understand how difficult it was to see such suffering.' His words calmed me and we turned on the television to find that my speech was being replayed. A note was pushed under the door, from Vincent Browne, then editor of the *Sunday Tribune*: 'You were brilliant.' It was surprisingly touching; it had been an emotional and difficult trip for all of us. And then, in private, the tears began to flow.

The next morning we flew from Nairobi to New York and I met with the secretary-general of the United Nations, Boutros Boutros-Ghali. He thanked me for going to Somalia as president of a European country, and for putting its plight to the world and drawing attention to the need for action from the international community.*

Two years later, when I was asked by the Irish aid agencies to draw attention to the genocidal killing in Rwanda, which had taken place in April 1994, it was more straightforward getting the government consent I needed to travel. The only concern – and it was a big

---

* Not long afterwards, US troops were sent to Somalia, but the mission was generally seen as a failure.

concern at the time – was my safety. It was relatively soon after the killing sprees, and pockets of the country were still considered unsafe.

The visit to Rwanda was really shocking. There was a smell of blood and indeed still splatters of blood in places where massacres had taken place. In the corner of a community hall was a pile of clothes, and of children's shoes (horribly reminiscent of the museum at Auschwitz albeit on a much smaller scale). Nick and I were shown a wooden stump, darkly stained, where people in their dozens had been beheaded.

In Kigali we stayed in the Hôtel des Milles Collines, which had no running water. A place where people had taken refuge, it was in a very stark state. Although the electricity was erratic, I remember watching a CNN programme announcing the first cease-fire by the IRA. I suddenly felt very far from home. I was struck by the irony of being in a place suffering from the aftermath of genocide, with high tension between two communities still palpable in the streets, and learning of a significant step on the road to peace on my own island. Ceasefires would be broken, and the road would be rocky, but the momentum towards a lasting peace was building.

Prior to our flight home we were to stay overnight in Kampala, and the president of Uganda, whom I didn't know, had invited us to dine with him. My first response was politely to decline the invitation; Rwanda had been exhausting, stressful, difficult. But the reply came that there would be no speeches; the president was offering the dinner as a friendly gesture in appreciation of our mission. So we decided we would accept, and that was my first encounter with President Yoweri Museveni. I found him impressive in his efforts to be innovative for his country, which had been shattered by earlier, awful dictatorships.

The following year, I was nominated by the Irish Government to represent Ireland at the fiftieth anniversary of the United Nations, a commemoration involving some 180 heads of state or government. With this in mind, I thought that we should go back to Rwanda and bring the post-genocidal problems of that country to the table as the United Nations celebrated. Bride and Peter negotiated with the Taoiseach's Office and the Department of Foreign Affairs, and they agreed.

So Nick and I, with our small party, returned to Rwanda in October 1995, and this time we also visited its second city, Butare. As we arrived by helicopter, we flew over one building, and I couldn't understand what I was seeing: there were men on the roof, men crammed together in the courtyard. I was told that it was the prison, and so overcrowded that inmates had opted to move onto the roof.

Our first stop was to visit the university. Everything had been destroyed; it was an opportunity to be supportive, and we responded positively to requests for books and materials from Ireland. The university was in the process of changing its teaching courses from French to English, because Vice President Kagame and the Tutsis he led had all grown up, basically, in Uganda, speaking English. The Hutus were French-speaking.

Bride went off to the prison in advance of us, and when she was informed that the chapel, which also housed prisoners, had not been cleaned, she told the authorities that I was a particularly religious person and would want to see the chapel! This started a flurry of cleaning up. When we arrived I met a representative from the Red Cross, a German woman, for whom my visit had meant that she got to have more access, and conditions had been somewhat, even if temporarily, alleviated. We walked through a narrow space between sweating, barely clothed prisoners on either side, and with very few guards to separate us from them. If the prisoners had wanted to cause trouble, we would have been dead. It was tense, but they stood there impassive and mainly silent.

That visit enabled me to bring the concerns of Rwanda back to the United Nations, and to its host nation, the United States. The formal speech at the UN was limited to seven minutes, but I had also been invited to speak at Yale. In New Haven, I quoted a passage from Derek Mahon's 'A Disused Shed in Co. Wexford':

> They are begging us, you see, in their wordless way,
> To do something, to speak on their behalf
> Or at least not to close the door again.

That caused my voice to break, momentarily, because it was bringing back such a memory. (And my voice still breaks even now when I recount the visit.)

When I delivered my formal speech at the UN building in New York Nick was up in the gallery sitting beside Mrs Boutros-Ghali and Mrs Leah Rabin. (Israeli prime minister Yitzhak Rabin was assassinated shortly afterwards.) Nick and Mrs Rabin had their watches out to time our speeches. Nick was pleased that my delivery, unrushed, was bang on the seven minutes and, thanks to the drafting skills of Noel Dorr of Foreign Affairs, that it also contained real bite. But contributions from people such as Yasser Arafat and Fidel Castro totally ignored the time limit and exceeded forty-five minutes and longer.

This was 1995. Some speculation had arisen as to whether I would be a candidate for secretary-general of the United Nations. This was not something I was seeking. I felt instinctively that the role would be too political for me: I was less skilful in the arts of compromise and diplomacy and preferred to speak truth to power. But it was flattering to have my name canvassed. One social diarist had begun, as a tease, to refer to me as 'Mary Mary Robinson'. Boutros Boutros-Ghali was seeking a second term, and I could discern a slightly combative attitude towards me during this trip.

The third visit I made to Rwanda as president of Ireland was in March 1997. I was one of two European women invited to a pan-African women's conference organised by Rwandan women – widows, most of them – and the government of Rwanda. Paul Kagame recognised that if it was going to be able to cope with its still very damaged infrastructure, with its huge prison population, with the poverty of its people, Rwanda had to give full opportunity to women. The government was considering changing the law so that women could own land; it was already changing the culture so that women could hold government positions, and encouraged, through a quota system, women getting into parliament. I remember being incredibly moved at that gathering of distinguished and activist women. These were exactly the right policies, and symbolic ones, that put Rwanda back on the map in Africa for reasons other than the conflict and genocidal killing. Each time we had gone we had seen improvements, not least in the Milles Collines hotel, now with running water, but also in the rebuilding of infrastructure and houses.

Arriving back to a press conference at Dublin Airport, I remember saying, 'I have seen the future of Africa and she works!' It is a phrase I have been able to use many times since.

One of the duties of president is of course to represent Ireland abroad at more ceremonial functions. My first state visit was to Portugal, in 1991, and it was an early opportunity to learn about the endless protocols that govern formal dinners and inspections of guards of honour. A highly detailed book was prepared for me that seemed to account for every minute of the trip.

Mário Soares, a hero of mine, was now president of Portugal, and would be my host. He had been a courageous lawyer and activist in the 1974 Carnation Revolution, had been imprisoned by Salazar's authoritarian regime and had prevailed to become the country's first elected civilian head of government. He and his wife, Maria, were helpful, gracious, and fascinating company. The language we had in common was French.

The mayor of Lisbon at the time was Jorge Sampaio, a lawyer I had known when he was a member of the European Commission of Human Rights;* we had dined together in Strasbourg with our mutual friend Michael O'Boyle. On one of the many inspections of guards of honour that I was required to carry out, Jorge and I were standing side by side under the hot Lisbon sun when a rather portly army officer, captain of the guard, marched up to us and stood to attention, puffing out his chest to give his salute, whereupon, with a distinct 'ping', the medal pinned to his uniform flew into the air and landed on the ground by our feet.

Irish cameras were following my every movement closely on this, my first state visit, and I was acutely aware of Jorge standing beside me and, at all costs, of not catching his eye or reacting, as a tear of suppressed laughter rolled down my cheek. As we moved to inspect

---

* Jorge Sampaio succeeded Mario Soares as President of Portugal in 1996.

the guard of honour, Jorge said to me out of the corner of his mouth, 'Did you ever think that you and I would end up doing this?'

Sometimes the visits had stirring moments. Having been involved in the Irish anti-Apartheid movement with Kader Asmal, and having as a senator joined AWEPA, the European parliamentary group fighting Apartheid, I was especially honoured in 1994 to represent Ireland at the inauguration of President Nelson Mandela.

The evening before, Nick and I sat at a table with Archbishop Desmond Tutu and his wife, Leah, at a formal dinner in Pretoria City Hall. We heard President de Klerk give his last speech and withdraw. In the pause before the president-elect, Mandela, arrived, Archbishop Tutu turned to me and said, 'Mary, I wonder if you can really understand how significant these changes are? This is my first time in the City Hall!' I looked in amazement at this Archbishop of Cape Town and Nobel Prize winner and then could only tease him. 'After all, isn't it only a few weeks since you cast your first vote?'

Everything about the inauguration was special; the taking of the oath by the country's beloved Madiba (Mandela's Xhosa clan name), the fly-past and salute that caused a huge, visceral roar from the crowd below, the rows of young South Africans of all races singing together as one.

I was delighted, two years later, to be invited to pay a state visit to South Africa and to be so warmly welcomed. Kader Asmal, now a cabinet minister, introduced me as I arrived to address the South African parliament, and in his enthusiasm used the words 'we Irish', evoking mirth among the parliamentarians present.

At the opening banquet later that evening, President Mandela displayed both his affection for Kader and his wonderful capacity to tease. 'Now, we know that President Mary Robinson had the honour and privilege of being lectured by Kader Asmal.' Pause. 'And we know that her husband, Mr Nicholas Robinson, had the honour and privilege of being lectured by Kader Asmal.' A longer pause. 'As we now in cabinet have the honour and privilege . . .' and his audience broke into loud laughter and applause before he could complete the sentence.

★   ★   ★

Before he was even elected president, Bill Clinton and his wife, Hillary, had Northern Ireland in their sights. As a presidential candidate Clinton made a commitment to send a peace envoy to Northern Ireland, and he came through with that in the person of Senator George Mitchell.

I had met George Mitchell several times when he was Senate majority leader, and I admired his precise command of language and self-deprecating sense of humour. Before travelling to Belfast in 1995, he came to lunch with Nick and me. He explained that his role would be to help organise an economic conference so as to convey a sense that there would be a peace dividend if progress were made, with economic investment coming into Northern Ireland. He estimated that his stay in Belfast would be less than three weeks, but shortly afterwards, his terms of reference were altered, and he was cast into a mediation role that would keep him focused on Northern Ireland for more than three years. His was a significant contribution, for which he has been honoured in both parts of Ireland.

I remember having two youth football teams from Northern Ireland come to visit Áras an Uachtaráin, one from the unionist community and one from the nationalist, with community leaders from both sides. As it happened, I had invited former US president Jimmy Carter (another hero of mine) and his wife, Rosalynn, to lunch as they were in Ireland at the time, and we decided to hold on to the football teams until Jimmy Carter arrived, as they wanted to meet him. He was fantastic with the kids, rose to the occasion, and asked them all the right questions.

With one of the teams was a man called Billy who, having been a hard man, had become a community youth worker. I remember going over to him as he was stood by the mantelpiece in the formal reception room and asking him how George Mitchell was doing, that he seemed to be making great progress. Billy smiled at me and shook his head in mock exasperation. 'Och, President, he listened us out'!

President Clinton played a key role in helping to find a solution to the Northern Ireland problem. He understood the need to change the dynamic from a win-lose situation to one that enabled people to come to the table without blame being apportioned. He embraced what were known as the McBride Principles, a kind of ethical code of conduct for US companies doing business in Northern Ireland to counteract discrimination. This brought much-needed economic investment, and helped create the understanding that there could be non-violent change that was effective.

In 1995, a year after the cease-fire, the Clintons travelled to Northern Ireland for an immense visit – one of the highlights of his presidency. Tens of thousands of citizens lined the streets in an emotional display, and Van Morrison sang, 'Days Like This'. President Clinton had utterly captured the people's imagination. In his spirit of embracing those who were indicating they would cease violence, he publicly shook hands with Gerry Adams on the Falls Road, notwithstanding criticism from the British government.

Following this visit to Northern Ireland, the president and Mrs Clinton travelled to Dublin and paid a courtesy call on me at Áras an Uachtaráin. We had met several times before, and because of our similar background as lawyers interested in human rights, we got on well. We had a private lunch, and I invited them both to plant a tree in the grounds of the Áras, as was common practice with visiting dignitaries. After the tree-planting ceremony, when we were walking back up the path towards the house, Clinton said to me, 'So *are* you going to come on a state visit to the US?' That was the first I had heard of such an idea. When I replied that I would very much like to, of course, he responded that the many hints his people had been dropping to my people had been getting no response. I explained to him that while I had travelled quite widely in the United States as president by that stage, the Irish Government would not favour my going to Washington DC. Washington was considered 'Taoiseach turf'. He gave me a knowing look and said to leave it with him.

That evening, he spoke to the thousands of people who had gathered in Dublin's city centre, at College Green, and received a rapturous reception from the Irish crowd. In the middle of his speech, he said, 'I'm looking forward to welcoming President Robinson shortly to Washington.' And that was it.

The state visit to the United States, which took place in June 1996, was incredibly enjoyable and a great opportunity to strengthen that particular bond Irish people have with the people of the United States. Nick and I were put up in Blair House, the official state guest house in Washington DC and treated to the full splendour of an American welcome, culminating in a huge state reception in the White House.

Not all of my visits abroad ran so smoothly or were treated so positively by the press. In March 1995, straight off the back of a nine-day visit to Japan to promote Irish trade, I undertook a trip to three South American countries: Argentina, Chile, and Brazil. The visit began in Argentina, a country in which some 300,000 people claim Irish descent. The Irish embassy in Buenos Aires received many requests from Irish-Argentinians for a visit from the Irish president, among them a group of Dominican nuns serving in a poor area outside Buenos Aires. President Carlos Menem – whose son had been tragically killed in a helicopter crash only a week before, a tragedy that cast an understandable shadow over the visit – was facing an election and blocked this request for a visit, as it might draw attention to the government's economic and social failures in the region. In fairness, I didn't press the point. But the disappointed nuns complained to Irish reporters covering the trip that I wouldn't meet the poor, and that became the story back home.

Travelling from Argentina to Chile, my suitcase went astray, and very unfairly, because it was not at all her fault, I gave out stick to Bride about this. I was overtired and perhaps underprepared, and losing my suitcase was the last straw. I can sometimes get terribly impatient; it is a bad trait of mine. So I snapped at Bride, and she in turn was cross with me, and entitled to be, and that set the mood. Then we learned to our dismay that the notorious General Pinochet would be attending the formal dinner in Santiago. As is the standard protocol for such events, those invited formed a reception line to shake hands with the guests of honour. So Pinochet took his place in the line and, rather than publicly snub him and make a big scene, I coldly shook his hand when he offered it. Though Bride had manoeuvred to ensure we wouldn't be photographed, our handshake became the story.

For the Brazil leg of the trip I had been advised by a local priest in Rio not to visit favelas (shanty towns), as this would require an advance visit by police for security purposes and would be detrimental to the favela dwellers. Again the story was put out back in Ireland that I refused to meet the poor.

And so a trip that was really quite successful in terms of improving relations between Ireland and the three South American countries, and that opened up new connections in trade, was marred by these incidents of negative publicity. Despite the warm reception we received in each country, from both governments and the wider community, I was disappointed that the coverage had focused on negative and not entirely accurate stories. I tried to be philosophical about it. After all, I received more than my share of good publicity. But it jarred with me, especially the accusation of my not wanting to visit the poor.

A strong theme of my presidency was inclusivity, bringing outsiders into the conversation and involving people who had not previously had access to their president. This list, it seemed to me, was rather long.

One simple way to open up the Áras came to us in the summer of the first year. A tradition had grown up of throwing a garden party for the Red Cross, which involved erecting a marquee in the formal front gardens, and we decided to expand that to a three-day event and invite a wide range of people to a series of garden parties, enticing entertainers and musicians such as Irish folk singer Christy Moore and his family to come play for the assembled guests.

Meanwhile, every week, groups and individuals were invited to be shown around the formal rooms and to learn the history of the house, which had been purchased for the Lord Lieutenant in 1782. It served as the vice-regal lodge until it became the official residence of the governor-general in 1922 and then home to my six predecessors and their families.

Áras an Uachtaráin was no longer a secluded, isolated place, and the enthusiasm of people visiting was growing enormously. When we moved in, the large basement was dank and full of old newspapers and discarded furniture. Nick felt it had great potential, and he, Bride, and I discussed the possibility of developing a visitor centre there. Eavan Boland's husband, writer Kevin Casey, took on the task of researching and developing the house's story, as a team of workers from the Office of Public Works began to transform the unused basement into a modern visitor centre. Nick took great interest in this process and encouraged the National Gallery to lend the Áras some fine works from its collection so that they could be admired by visitors.

It gave me great pleasure to invite former Irish president Paddy Hillery, his wife Maeve, and his family, and Rita Childers, the widow of former Irish president Erskine Childers and their daughter Nessa, as well as Eavan and Kevin and some other close friends, to the opening of the centre. Now when groups came to the Áras they would begin their tour in the visitor centre before coming up to the formal reception rooms where, if available, I would greet them and offer refreshments. The house was bright and open, the staff welcomed everyone warmly. Nick described such improvements as 'invisible mending'. I felt he played a significant but largely invisible role in achieving that mending.

The light in the window of the Áras was not just for those who had emigrated. It was also for those who felt marginalised or excluded within the country. I remember some of these visits vividly because of the underlying emotion that affected us all. It was a particular pleasure, for example, at an early stage, to welcome members of the Irish gay and lesbian community to the Áras with one of their leaders, my friend Senator David Norris. We chatted and took tea together, and there were short speeches marking the occasion as of deep significance to them. Then I suggested we go outside and down the steps under the portico to view the light in the window, so that photographs could be taken. Immediately there was a nervous reaction from several people, who moved back instinctively. It became clear that at least half of those present (these leaders of the Irish movement) were not 'out' to some individuals, be it a family member or perhaps an employer, and so did not wish to be photographed with the president in this capacity. This, more than anything that was said, brought home the pain of the group's exclusion and stigma.

Another group I was keen to include was the travelling community, whose cause I had supported as a lawyer. I visited various projects run by and on behalf of travellers and invited some groups to the Áras. While careful not to encroach upon the political arena in what I said, I did initiate an award for design of traveller accommodation and held a reception in the Áras for those participating.

When I hosted a special reception for groups representing those who were unemployed, their spokesperson's words in thanking me stayed in my mind long afterwards: 'The unemployed don't get many invitations.'

When I was elected as president, I knew I needed to exercise the role with confidence *as a woman*. I recognised it as being a positive factor, as something I could use to open up the national conversation about who we, the Irish, were. Because I was the first woman president, I could be innovative, could use my imagination, could put an emphasis on symbols: the light in the window, the hand of friendship to both communities in Northern Ireland. Being a woman president allowed a more nurturing role, giving further emphasis to caring about those involved in local community development. It allowed me to fulfil what I had said in my inauguration speech: 'As a woman, I want women who have felt themselves outside history to be written back into history, in the words of Eavan Boland, "finding a voice where they found a vision".'

Some women I approached would say, 'Oh, don't bother about me. I'm only a housewife.' That they should feel that way raised a question about the priorities of the women's movement in Ireland, and caused me to rethink and foster an inclusive approach, supportive of all women and respectful of their lives, and their choices. I was always looking for occasions to recognise and value contributions that had been overlooked.

The Irish Countrywomen's Association, for example, impressed me with the range of its activities, even though in the past I might have thought of it as being an old-fashioned rural women's association unconnected to what I was doing. I had come to realise that these were salt-of-the-earth people, the backbone of the country, really making a difference in their communities. Equally, women's groups in urban settings, such as the Irish Housewives Association, and smaller groups, particularly in inner city areas, were applying their strength, commitment, and energy in addressing the tough circumstances in which they found themselves by making things better for the elderly, for those with disabilities, and so on. This was mirrored in the heroic Northern Irish women's groups I met, even more impressive in the context of violent sectarian conflict. It was predominantly the women who came out from their conflicting neighbourhoods and housing estates to build links across the divide who brought the beginnings of reconciliation.

On the international level I took part in a meeting in Stockholm in 1996 where a number of women heads of state and government decided to form the Council of Women World Leaders, initially chaired by Iceland's president Vigdis Finnbogadóttir. At that time there were only half a dozen members; now there are more than forty. (Following Vigdis, then Kim Campbell, former prime minister of Canada, I went on to chair the council for six years, from 2003 to 2009 and handed over to Finland's president, Tarja Halonen.) One objective was to bring together women ministers with different portfolios, such as environment or finance, to stress the gender dimension of their work; another was to enhance the visibility of women at the highest level in order to offer encouragement and precedents for potential women candidates. I remembered that when I was campaigning for the Irish presidency, I was inspired by the fact that there was already one directly elected woman president in Europe, Vigdis. It could be done!

But as a woman president, I still encountered entrenched sexist attitudes. In 1997, I was invited to an event in Rome to celebrate International Women's Day, on 8 March. The Irish ambassador to the Vatican at the time arranged an audience with Pope John Paul II. As soon as the proposed visit was announced, discussion began as to what I would wear. Traditionally, a Catholic woman would wear a penitent's black dress and mantilla – (a black veil to cover the head) when in the presence of the Pope. But I was not meeting the Pope as a penitent Catholic woman, but as a head of state actively seeking to promote a tolerant, inclusive Ireland. To wear a mantilla would, in my view, send out the wrong signals. I wore a perfectly sober, dark green outfit and was accompanied by Bride, who wore an equally sober outfit in navy blue. However, given the day, we both wore, pinned to our coats, the symbol of International Women's Day: a sprig of yellow mimosa.

We were met by looks of horror from some Irish officials as we emerged from our hotel, bare headed, and by an audible collective gasp from the honorary committee – a dozen Italian men in smart formal dress with the vivid silk sashes and elaborate head-gear of their orders and decorations – who had gathered to greet us at the Vatican as we got out of our car. I met with the Pope, in private audience. He was a remarkable man and a remarkable pope. He greeted me warmly; he had quite a twinkle in his eye and knew well what I was about. We talked about Ireland, we talked about human rights, we talked about Rwanda and Somalia, the work of the Pontifical Council for Justice and Peace – serious discussions into which my choice of clothing did not enter.

12

Neither Fish Nor Fowl

We can best help you prevent war, not by repeating your words and repeating your methods, but by finding new words and creating new methods.

Virginia Woolf, *Three Guineas*

IN EARLY 1997 I had to make a decision as to whether to stand for a second seven-year term as president. This was an agonising decision, one of the most difficult of my life.

If I were to run for a second term, I would very likely be re-elected. I could nominate myself – as provided for in the Irish Constitution – and would almost certainly run unopposed: the Irish people had by now thoroughly identified me with the post, and for my part, I loved the work and looked forward to each day serving as president to the best of my ability. I could very much see the attraction of continuing, particularly in order to support reconciliation and further contacts between communities in Northern Ireland, and North and South, to complement the political peace process as it got under way. But I felt my major goal had been achieved. I had shown that the office of president was a vital part of our Constitution and could fulfil important roles internationally, nationally, and locally. Inevitably the role involves repetition on key national occasions. Could I do this with the same enthusiasm and commitment in a second term? Yes, for another three or four years. But another seven years? The premise of my running for president back in 1990, and throughout my term, was of opening up the presidency, reshaping it into a dynamic office. If I had achieved something close to that, would my ongoing presence in the position not result in stagnation?

This was the view that Bride took, encouraging me to step down, as did Nick and my wider family and friends. They felt that now was the time to have an election and let a successor continue the work.

As I recall, the only close friend who encouraged me to stay on was Eavan Boland, who wanted me to consolidate what had been achieved. I did want to remain in office, but I had doubts about whether I could do it for another seven years with the same 100 per cent commitment.

There was, of course, the question of what I would do next, if I did not put myself forward for a second term. I was still only in my early fifties. Returning to the Law Library to practise law was out of the question. In truth, working in Ireland in any capacity might be difficult, certainly for the first few years, because of the high profile I had had as president. It seemed better to look abroad. A university professorship was a possibility.

I was still reluctant to make a decision, and Nick and I took a brief holiday in Malta in February 1997 to make up our minds. Bride had strongly advised that if I chose not to run I would have to announce this early enough in the year to enable new candidates to declare themselves and an election to take place. Nick and I walked for miles, arguing out the pros and cons. We came home with no firm decision, but I began to see that it was better – for the office of president, for the Irish people, for me – not to seek a second term.

Having discussed the matter briefly with my family and close friends, I publicly announced my decision in March; the reaction came as a complete surprise. There was so much coverage – eulogising – in the Irish media, print, radio, and television, emotional appeals from many quarters that I change my mind, that it felt almost as if I had died! I telephoned my father, my brothers, my children, simply to stave off this sudden lonely feeling that the coverage invoked.

But within a few days, as is the way of these things, attention turned to who would be my successor, as candidates began to declare. And it was satisfying that four of the five candidates who did eventually declare were women, one of whom, Mary McAleese, succeeded me.

I heard shortly after my announcement – and the news came like a jolt of electricity – that the first UN High Commissioner for Human Rights, José Ayala-Lasso, had resigned unexpectedly, in the middle of his first term, to return to his native Ecuador and take up the post of foreign minister. The Geneva-based office was still in its infancy. It had been established by the UN General Assembly in December 1993, following the world conference on human rights in Vienna earlier that year, to the surprise of the many who considered there would be too many governments in opposition to the move, fearing scrutiny and interference. I was excited at the thought of putting myself forward for the newly vacant position: this type of work went to the heart of what I was interested in and had worked on at the national and European level, and offered me a chance to get back to an activist role, this time on the world stage.

I sought private advice from the High Court judge, Mella Carroll, who I knew travelled frequently to Geneva for International Labour Organization duties. She was unenthusiastic, telling me, 'The office is small and underfunded.' Peter Sutherland, a former Law Library colleague and EC commissioner, who had headed the General Agreement on Tariffs and Trade GATT talks in Geneva, was even less encouraging, advising me not to seek the post, that it was contentious and disorganised. But it seemed to me so serendipitous that the position had unexpectedly become available that my heart was becoming set on it, and perhaps I brushed off too lightly the underlying warnings of these colleagues. Even when I approached the Taoiseach, John Bruton, to inform him that I was interested in the position, he initially expressed doubts, telling me that he had heard rumours that José Ayala-Lasso had left because of rifts and lack of resources. When I pressed him, and he saw how keen I was, he relented, and a committee tasked by the government then set about running a vigorous global campaign to secure the position for me, sending formal letters to governments seeking their support for my candidacy – it had become a competitive race as a number of countries had put forward candidates – and putting my credentials to the UN permanent representatives in Geneva and New York. The post, at the level of under-secretary-general, was the highest in the United Nations sought by an Irish candidate.

When Kofi Annan, who had replaced Boutros Boutros-Ghali as secretary-general of the United Nations in January 1997, decided that I was the candidate he wanted to place before the UN General Assembly, he put severe pressure on me, during a meeting in New

York in July, to come to the post almost immediately. Explaining that the office had been without either a High Commissioner or deputy for several months and that morale was at a low ebb, he emphasised the urgency of being in the post before the General Assembly met in New York in the third week of September. I felt a double pressure: that there was clearly an urgent need to fill the post, to give leadership on human rights, and furthermore, if I was not available until after 3 December, – when my term as president was officially due to end, – Kofi Annan might well choose to put forward another candidate.

Greatly torn, I reluctantly agreed to leave the presidency about ten weeks before completion of my term to start the new job on 12 September. It was ill-judged on my part, a serious miscalculation. Now I know that UN offices are often in crisis, and that Kofi Annan probably would have waited for me until my term ended if I had insisted; and even had he not, it wouldn't have been the end of the world. I wish I had insisted on serving out my full term as president. It did not occur to me at that time – a vigorous election campaign was in full swing, and I was becoming yesterday's news – that I might leave some people in Ireland with the impression that the presidency was less important to me than a UN job. That was never the case. The honour of serving as Irish president was, and will always remain, the greatest of my life.

On 12 September 1997, I fulfilled a last engagement as president, opening a housing project for Focus Ireland, an event organised by my friend Sister Stan. I was moved by the large crowd and the warmth; I felt very torn. Nick had come back from Geneva, where he had settled Aubrey, our youngest, now sixteen, into a rented house and an international school. Tessa had an apartment in Dublin and was continuing her studies, and William was still in Glasgow studying architecture at the University of Strathclyde. Nick understood the emotional turmoil I was feeling, and tried to lighten the mood, but for once I couldn't laugh, though I appreciated his efforts.

Nick and I returned to the Áras for a last time, and in a simple ceremony, I signed my letter of resignation and spoke briefly, thanking Peter Ryan and his colleagues who had served me with such dedication over the seven years. Some were in tears; I was close to tears myself.

I had dressed in a dark business suit and carried a briefcase, signalling that later that day I would make the transition from being president to assuming my new duties as High Commissioner. Leaving the Áras, I pinned a broad, fixed smile on my face for the cameras, afraid that at any moment I would betray my emotions. We waved goodbye, drove to the airport and caught a flight to Geneva.

On the aeroplane, Bride Rosney – who, at great inconvenience to herself, had agreed to join me in Geneva for the first eleven months to assist with the transition – tapped me on the shoulder and said, 'It is noon in Geneva. Congratulations, you are now the UN High Commissioner for Human Rights!' Without her at my side, I doubt I would have been able to stay the course.

We were brought to the huge Palais des Nations, Geneva's UN building, up to the fifth floor, down long corridors, and to a far corner where the High Commissioner's office was located. Although it was fine and well furnished, its obscure location gave me a sense of the lowly status of the new office in the UN, and this was further emphasised when I discovered that the other staff members' offices were scattered about over a number of floors, down endless, gloomy corridors. (Picture a 1950s hospital.)

Ralph Zacklin, a UK national who had served the United Nations secretariat since 1973 and was now assistant secretary-general for legal affairs based in New York, had been standing in as acting head of the office since March. He had travelled to Dublin to brief me prior to my arrival. Those briefings had presented a picture of an efficient, well-organised office, but as he greeted me in Geneva, Ralph informed me that the full staff was assembled to meet me, and warned me, discreetly, that morale was low and that I should try to give them encouragement. I looked at him and had a sudden realisation that I was facing a much more difficult task than he had initially led me to expect.

The staff of about 150 had gathered in palpable, apprehensive silence. I spoke of the honour of being appointed and said that our role was a significant one. The secretary-general's reform package, passed three months previously, meant that there would be a single Office of the United Nations High Commissioner for Human Rights (OHCHR), incorporating the Centre for Human Rights that had been tasked previously with the human rights portfolio. We were one team now and should work as one team. In the following days, I began to see the depth of resistance to this idea, and the challenges of reforming large bureaucracies. I met many new colleagues who liked to refer to themselves as officers of the 'Centre' for human rights, as if that were somehow more prestigious than being part of OHCHR. I would soon discover that the first High Commissioner and the head of the then-Centre for Human Rights had essentially been at war with each other, and fighting over scarce resources. As a small way of beginning to move the office forward and build a new sense of teamwork among senior management, I put in place a nominal fine during senior staff meetings, charging anyone who referred to the 'Centre' (though, of course, I never exacted any real fines). It took a while, but in the end, people did start to identify with a single office for human rights.

Returning to our rented home that first evening in Geneva, Nick, Aubrey, and I decided to try out a restaurant in our neighbourhood, the Café du Soleil, for supper. Sitting in the open air, I found it strange to be out in public without any security, or any element of protocol. As we sat there together, I suddenly felt waves of tiredness washing over me, partly the strain of that final period in Ireland, with no break before plunging into my new post, and partly the realisation of how much work lay ahead – of which I had had but a glimpse that afternoon.

Every person I met that first week – every expert on the various committees that monitored governments' human rights performance, every special rapporteur* who investigated rights abuses around the world – complained bitterly about the lack of resources for their vital work and the lack of support for what they were

---

* Special rapporteurs are individuals who work voluntarily on behalf of the UN with specific mandates, whether thematic or by country, to monitor and investigate human rights problems.

doing: somehow I was being looked to to put this right. There were so many problems; it was like lifting stones and each time finding more creepy crawlies underneath, scurrying away to hide. And the way to tackle them was not clear to me. I began to come into the office earlier and earlier each morning, and leave later every evening.

At the end of the first week, Bride and I travelled to New York for the annual UN General Assembly session. It was the first occasion for me to meet senior UN officials, foreign ministers and ambassadors who would be among my key interlocutors, and an early opportunity to discuss the work of my office and its importance for the UN system. I stayed at a welcoming Irish hotel, the Fitzpatrick Manhattan, where I had been offered a special rate to stay in their 'Mary Robinson' presidential suite! It was to become my New York home.

Bride and I, steeling each other with courage, arrived at the UN building in New York for the first time and headed for our offices on the twenty-ninth floor. We entered with no fanfare or welcome, found an elevator, and got in. Nothing happened. We stood there for quite a while before Bride nudged me and said, 'For the elevator to work, you have to press the button!' It had been seven years since I had pressed an elevator button. There was a lot of that: re-learning how to behave, how to wave down a taxi, buy a subway ticket, remember to bring money with me, use a credit card, buy basic foodstuffs.

We found the small liaison office chaotic, with papers everywhere. Bride even found an uncashed cheque, payable to the office, in a desk drawer. It was becoming clear that as well as giving leadership in the UN system on human rights, supporting the special rapporteurs and the UN treaty bodies, and being close to the victims of violations, I would have to restructure the office, sort out the contractual positions of staff members and build up morale. I had not anticipated that basic systems of management would have to be put in place on top of the major substantive human rights work. It was a comfort to have Bride at my side, but the task was formidable.

During that General Assembly meeting in September 1997 I met with Kofi Annan for the first time since taking up office. We had a friendly meeting – I was still, a bit, the golden girl – and discussed his new reform package and the restructuring associated with it. As a new secretary-general, Kofi had seen a need for greater coordination of the various agencies in what he identified as four core areas: peace and security, economic and social affairs, development cooperation, and humanitarian affairs. As part of this institutional reform, he formed executive committees as clusters of agencies into those four areas. The new OHCHR would be a member of all four executive committees as part of the stated goal of 'mainstreaming human rights'. I remember asking him how he envisaged that it should work, and his answer, accompanied by a smile: 'That's for you to figure out, Mary.'

Also scheduled for that New York trip was a series of introductory meetings with various foreign ministers. For each, I had been given background material and issues I should raise with him or her, which I duly did, ruffling feathers along the way. For example, when I met with the Algerian foreign minister and raised the relevant points – ongoing conflict and the widespread and brutal massacre of civilians without investigation by the Algerian authorities – he visibly bristled, countered very forcibly, and then issued a statement complaining about me. I realised I may have gone in a bit too bluntly with him (for his liking anyway). It was a lesson for me: this was going to be tough work, and I needed to hone skills that were a mix of diplomacy and firmness.

A week later we returned to Geneva where the level of problems within the office was becoming more and more apparent. Upon my appointment the workload for my staff had increased significantly as I tried to find out what the issues were. Some of the staff really felt the pressure. My executive assistant had known the demands of her job before I took over, but the additional strains proved too much, and fairly early on she took sick leave for work-related stress. I was blessed, then, to find an extremely competent and resourceful replacement, Cecilia Canessa, who was in the office as second assistant. Cec, from Uruguay, was to become a great friend.

I became aware that contributing to the stress in the office was the fact that staff were on different types of contracts. Some held long-term contracts and had impenetrable job security; others, many of whom were highly capable and coping with huge workloads, had contracts as short as one month at a time, which made their work and personal lives, often far from their home countries, extremely difficult to manage. I met many young staff members who were eager, but constantly worried about their next contract. Others had been in the system for many years and now earned good salaries with generous UN benefits, but seemed to have lost much of their drive to do their best in their work. Both kinds of situations were unsatisfactory.

As part of the compromise of establishing a human rights office with a mandate to address state performance on some of the most sensitive issues, the office was part of the general secretariat controlled out of New York, rather than being an autonomous UN body, such as UNICEF, for example, where all management and staff issues were independently controlled. Every March, the office reported on its work to the Commission on Human Rights (since replaced by the Human Rights Council), and every September, to the General Assembly. This arrangement allowed for an intrusive level of micromanaging by member states, who sought to limit the power and influence of the office. The creation of new posts required intergovernmental approval, and even the most straightforward recruiting or other staffing decisions could often take nearly a year to find their way through the UN bureaucracy.

And on top of all the management hurdles, and despite all the rhetoric on human rights, the budget for the office was less than 2 per cent of the United Nations' entire operating budget. This meant that in order to be effective, we would have to seek 'extrabudgetary funds' from philanthropic foundations and from donor countries. Thus I could find myself, in a meeting with a foreign minister, on the one hand more or less begging for funding, and on the other raising sensitive, often unwelcome issues around human rights concerns. I was careful from the outset to seek funding from smaller countries around the globe as well as from the major economies, even if in modest amounts, to avoid accusations of a Western, donor-driven agenda. Out of any funds raised by my office in this way, the secretariat in New York retained 13 per cent.

As I gradually settled into my role, I was of course expected to make numerous public statements each week — to the media, in front of government delegations, at meetings of non-governmental organisations, and before various UN gatherings. I became increasingly frustrated by the pro forma draft texts I was given for speeches, which were completely without life. Many of my colleagues in the office, as is typical of the international civil service, were accustomed to preparing bland speeches designed to avoid any possible controversy — 'I open the committee on . . . I urge you to do your work well.' Often I would be given the draft first thing in the morning for a function later that day and would hastily have to try to inject some energy into it. One of the dullest was a text I was given to open a meeting of a UN working group preparing a declaration on the rights of indigenous peoples. Overly formalistic and full of acronyms I didn't know, it was a text I just couldn't deliver. Instead I went in, introduced myself and then told them a story I had learned during my presidency, of how, in 1847, the Choctaw nation in the United States had donated $173 to a famine-stricken island thousands of miles away, Ireland. I had taken the opportunity, 150 years later, to travel to Oklahoma to thank them: we should never forget that people remember the grievances of history, but they must also remember those who reach out in such generosity. That story was my way of making the point to governments, and the experts gathered, that their work needed to speak to the legacies of past injustice but also point the way forward. Telling stories had been a prominent tool in my kit as president, and I intended to go on using stories to try to motivate and lead as High Commissioner.

I would soon discover that my words as High Commissioner were studied and widely repeated, so I needed to use important occasions to deliver messages effectively. As I got to know my senior colleagues, it became clear that my first spokesperson, John Mills, an Australian whose natural integrity shone through, would be a key ally in helping me craft my early statements. So, too, would Tom McCarthy, who had an 'institutional memory' of how the human rights system had developed, and later Scott Jerbi, who was to become one of my closest confidantes in the office.

Meanwhile, Nick was having a tough time. I was so preoccupied with the work and commitments that there was no question of balancing family life. We had rented a modest, urban house with a small garden. Aubrey was attending the international school, which he seemed to be enjoying. During the few weeks I wasn't travelling, I left the house before 7.00 a.m. and was rarely home before 7.00 p.m. At least once a week I would attend a formal evening function in Geneva, hosted by an ambassador or other dignitary, or speak at a civil society function, where almost everyone else would eat heartily and I would spend most of the time speaking, answering questions, and trying to eat as quickly as I could when I got a moment. (It reminded me of the speed with which my father and two older brothers, doctors on call, ate their meals.)

For the first month, Bride stayed with us, until she found an apartment of her own. We would arrive home and then talk shop for another hour or so. Nick was left to do all the shopping, look after Aubrey to the extent that a teenage son would allow, and prepare the evening meal. This was in stark contrast to the role he had played as my trusted confidant when I was president, attending functions and travelling with me. There was no question of Nick accompanying me when I travelled as part of my work; he was not encouraged to play any spousal role whatsoever within the United Nations, which came as quite a shock to the system: during the presidency we had worked as a team. Now I would sorely miss using him as a sounding board, his support, advice, and the way he could relax me by finding the funny moments. He formed a dim, – but accurate, – view of how the United Nations treated spouses.

From relatively early on in the job, I had trouble sleeping. For the first time ever I began to take heavy sleeping pills. They only compounded the problem, being not particularly effective at night and dulling my senses during the day. The responsibility of becoming aware of human rights violations in different parts of the world – reported to us by special rapporteurs and by organisations such as Amnesty International, Human Rights Watch, and many smaller local groups – weighed on me. How were we going to be effective in responding? I would lie awake and turn nightmarish problems – reports of massacres, rape, and sexual violence – over in my mind.

Gradually I began to get a handle on some of the key problems facing the office and, with Bride's help, started to plan, started not just to learn and do but to step back a little and think. I decided to use an invitation to give the Romanes Lecture at Oxford University on 11 November 1997 to set out my thoughts on my first months in the post and the challenges that needed to be confronted not just by my office but by the United Nations as a whole. Knowing this would be a key speech, I asked John Mills to work with me on it. We had already become good friends; I recognised his journalistic skills and even more his passion for human rights.*

Professor Philip Alston, in an article in the *European Journal of International Law*, described the role of the UN High Commissioner for Human Rights as 'neither fish nor fowl'. It mattered a great deal to me that the agenda of human rights embrace more than civil and political rights. In addition to a prohibition on torture, the right to a fair trial, freedoms of speech, of the press and of religion, the international human rights agenda must also include economic, social and cultural rights: to food, safe water, health and sanitation, and education. Yet these rights were often marginalised and seen by many Western countries as merely aspirational. I advanced the argument that extreme poverty is itself a human rights violation, that eliminating it was an essential part of a state's obligation to its citizens.

I appreciated the opportunity afforded, as part of the Secretary-General's reform package, to 'mainstream' human rights into all facets of the UN programmes. Our task under this mandate was to make human rights a priority, a 'core purpose' of the United Nations, and maintain a high profile at every level of endeavour – from humanitarian work and peacekeeping to development. But despite the key role that Kofi Annan had given to the High Commissioner, I had observed by now that if I were not present or represented at committee meetings, the topic of human rights would not be raised, and even where I was represented, it was often a struggle to urge a human rights perspective.

---

* Sadly John Mills died after a short illness in 2001 at age 46.

I did not want to pull my punches. I had in mind the inscription that Seamus Heaney had written for me on a specially bound poem he arranged to be sent to the airport for me as I was leaving Ireland on 12 September: 'take hold of it boldly and duly'. At one point, drafting the speech, I turned to John Mills and, referring to my criticism of the UN leadership's attitude towards human rights, asked, 'Can I say that?' His response was, 'You can, Mary, but is it wise?' One of my traits, and it has served me well even if it has put my career on the line on occasion, is to stay true to my beliefs. So at the Romanes Lecture at Oxford I spoke my mind:

> We still have widespread discrimination on the basis of gender, ethnicity, religious belief or sexual orientation, and there is still genocide – twice in this decade alone. There are forty-eight countries with more than one fifth of the population living in what we have grown used to calling 'absolute poverty'. This is a failure of implementation on a scale which shames us all. So much effort, money and hopes have produced such modest results. It is no longer enough to hide behind the impact of the cold war and other factors limiting international action in the past. It's time instead for a lessons-learned exercise . . .
>
> Almost by definition and certainly according to its Charter, the United Nations exists to promote human rights. Somewhere along the way, many in the United Nations have lost the plot and allowed their work to answer to other imperatives. This is the root cause of much of the criticism that is levelled at the organisation – you hear it couched in terms of complacency, of bureaucracy, of being out of touch and, certainly, of being resistant to change.

After I spoke, the chancellor of Oxford University, Roy Jenkins, who clearly foresaw the reaction my words would provoke, commended my speech as being 'very brave'.

The speech shook the United Nations. It was quickly boiled down to the High Commissioner stating that 'the UN has lost the plot!' Next day, I received a telephone call from Kofi Annan saying that I needed to remember that as a staff member of the UN, I owed it, and him, full loyalty; that I should be careful about making criticisms publicly that could be used by critics of the United Nations. This came as a surprise to me, and concerned me a great deal: that Kofi believed I was somehow being disloyal. I reminded him, in turn, that he had encouraged me to 'stay an outsider' as long as I could in this job, that my speech had been a rallying call advocating a strong UN. As it turned out, shortly afterwards the UN department of public information adopted and published the full text of the speech on its website.

A key function of the office throughout 1998 would be to mark the occasion of the fiftieth anniversary of the core human rights document, the Universal Declaration of Human Rights. The year-long celebrations were to begin on 10 December 1997, and it seemed fitting to me that we would launch the anniversary year in South Africa, where President Nelson Mandela was due simultaneously to launch South Africa's plan of action for human rights.

We constructed a trip – my first major field trip as High Commissioner – around this, planning to visit Uganda, Rwanda, and then South Africa.

Bride Rosney, John Mills, and I travelled first to Uganda, where I met with President Museveni. At this stage, following our meetings after trips I had made as president of Ireland to Rwanda, we had a good relationship. I raised concerns about human rights issues in Uganda first, based on the briefing provided for me by my office, then I sought his advice about what were ongoing concerns over Rwanda's activities in the Democratic Republic of Congo. Rwandan forces had entered the country, nominally to capture perpetrators of genocide, and were reportedly terrorising local villagers and pillaging diamonds and other resources. In addition, Rwanda, as I well knew, had an overcrowded and unsustainable prison population. Museveni, in his philosophical way, discussed these matters and told me of a local practice, *gacaca*, a traditional method of dispute settling, presided over by elders of a village. He suggested that this might be applied successfully in Rwanda to reduce the backlog in processing prisoners.

Having travelled to Rwanda on three occasions as president of Ireland, I felt a warmth towards the people of that country, which I supposed, was likely to be reciprocated. So I was taken aback at the blatant hostility directed at me now that I was a UN official. Bride requested a meeting with Paul Kagame (Rwanda's vice president at the time but known by all to be the real leader), and it took much back-and-forth before Kagame, together with some members of the government, agreed to meet us.

Their attitude was suspicious and critical. I got the sense of a dismissive disregard for my office, and for the United Nations generally. In many ways, of course, this was understandable, given the failure of the UN at the time of the genocidal killing in April 1994. I was mistaken in thinking that my personal record would make a difference – the hurt and resentment ran far too deep. When I began to raise concerns, the response, forceful and aggressive, was that I did not understand Rwanda. I struggled to communicate with Kagame on a human level, reminding him that he had visited Ireland while I was president, that I had driven to the town of Strokestown to meet him at the famine museum there. But, unresponsive, he shrugged these efforts off. My confidence was rapidly diminishing. I brought up the question of the prison population and Museveni's suggestion of *gacaca* courts, but mispronounced the word, so that Kagame and his men all laughed at me in a humiliating way. (Within a year, however, the Rwandans would establish *gacaca* courts to tackle the processing of the prison population.)

I pressed on, raising the concerns about the activities of Rwandan forces in the Democratic Republic of Congo. At that point the vice president stood up, told me angrily that he would not discuss this with any UN official, and that our conversation was at an end. I left the meeting bewildered and frustrated.

That night, staying in the Milles Collines hotel, with mosquitoes whining around the room, I barely slept.

At a press conference on our departure for South Africa, I spoke candidly and bluntly of serious human rights violations: arbitrary arrests, prolonged arbitrary detentions, and serious overcrowding resulting in inhumane conditions of detention. I spoke as a friend of

Rwanda, but my words revealed the frustrations I was feeling following the meeting with Kagame. I just didn't have the capacity to be as balanced and diplomatic as I might have been. For their part, Rwandan authorities issued an immediate retort, virtually cutting all ties with the UN.

As we flew from Rwanda to South Africa my mind was racing. I felt stressed and unhappy with how the press conference had gone. Was I really up for this job? Was I going to be able to fulfil this role? I replayed the trip to Rwanda in my mind. I tried to hold it all in but couldn't help the tears that began to well up. I could see the looks of concern on Bride and John's faces but brushed off any attempts to communicate. I was confident that my criticisms of Rwanda were warranted but I still had my doubts: Had I simply parachuted in, a Westerner pointing the finger without fully understanding the traumas of a country post-genocide?

Landing in South Africa, I soon saw that my statement in Rwanda had caused a stir. A certain tension was in the air. The prevailing word had it that I had spoken bluntly and, it was alleged, unfairly. On the other hand, in meetings with some of the human rights organisations the feedback was positive; that I had taken a strong stand. I received a personal message from the UN High Commissioner for Refugees, Sadako Ogata, thanking me for speaking truth to power, as so few people were prepared to do. At a lunch in my honour, the speaker of the South African parliament, Frene Ginwala, said to me, 'Didn't you know that it is an impossible job? Never meant to be carried through.' I was beginning to get a sense of that.

I was, nonetheless, greeted warmly in South Africa and felt gratified standing shoulder to shoulder with President Mandela, the legendary Madiba, as he launched both the fiftieth anniversary commemoration of the Universal Declaration and the newly published South African plan of action on human rights. At one stage he turned to me and said quietly, 'You know, Mary, there were people in the past who called me a terrorist.'

When I came back from the long trip to Uganda, Rwanda, and South Africa, I felt exhausted and depressed. I didn't want to ease up at weekends. All I wanted to do was work; there was so much for me to sort out. I remember at one stage Aubrey pleading with me to come for a day to Lausanne to relax. I agreed, and the three of us, Nick, Aubrey, and I, took the train from Geneva to Lausanne and strolled, sightseeing. I couldn't unwind. After lunch Aubrey tried to persuade us to walk around the lake but I didn't want to. I said that I wanted to go home; I had things to do. Back in Geneva, Aubrey wanted to show me an exhibition he'd been to the previous week. I didn't want to go. My energy and thoughts were elsewhere.

I returned to our home in the West of Ireland for the Christmas holiday. I was exhausted, and looked it. I had pushed my stamina to the limit and found that my usually reliable powers of recuperation were slow to come. Not having taken any break between jobs was adding to the toll. I found it difficult to relax or engage with my family. I had no appetite and absented myself from some family meals, preferring to be alone, trying to figure out how I was going to cope with the year ahead. I remember attending Mass on Christmas Eve in our local village. Formerly, as president, I had gone up to the front row; now I slipped in to the sixth or seventh row. I felt a lack of the previous warmth; it may have been in my imagination.

My eldest brother, Ollie, was home from New Zealand with his family that Christmas, and Henry and Adrian were there, and my father. They all expressed concerns about my health, Ollie most vociferously, warning me that I was straying into nervous breakdown territory and needed somehow to pull myself together. Ollie's bluntness hit home, drawing a quiet reaction of sibling bloody-mindedness, and bit by bit, I started to recover.

Over the next ten days or so I spent a lot of time outdoors in the fresh air, going for long walks. The permanence of the natural surroundings, the lake, the mountains, grounded me. I had time to reflect in peace, away from the daily task of waking up every morning to take on the burdens of the world. I began to brighten, and become more like my old self. Nick came with me on some of these walks and helped me pull myself together. 'You're not one to give up, Mary. You can do this job as well as anyone, and it's a job that needs doing.' I was grateful he didn't suggest that I give it up.

13

Boldly and Duly

We are not permitted to choose the frame of our destiny. But what we put into it is ours.

Dag Hammarskjöld, *Markings*

BACK IN Geneva, I dug deep and found my resolve. I ditched the sleeping tablets and fell back into a regular sleeping pattern. My sense of purpose deepened. I was going to work, with my colleagues, to bear witness to violations of human rights; to shine a light on problems the world wanted to forget; and to demand accountability – from states in particular, but also from the United Nations itself and, where appropriate, from corporations. We would also strive to provide an effective secretariat and support to the Commission on Human Rights, to the various UN committees that held governments accountable, and to the UN special rapporteurs, the frontline human rights defenders. My resolve was to give leadership: to use whatever tools I might have to bring home the need to prevent serious human rights abuses, to hold perpetrators to account, and to deepen the culture and understanding of human rights worldwide.*

The fiftieth anniversary year of the Universal Declaration of Human Rights was an ideal occasion to bring human rights to the fore globally, an opportunity to make them relevant to people's lives, as part of their birthright. I remembered how inspired I had been, sitting in my school library in Mount Anville, reading about Eleanor Roosevelt and her vision of why the world needed a common set of values. The Declaration ranks as one of the great aspirational

---

* For a more detailed account, see Kevin Boyle, ed., *Mary Robinson: A Voice for Human Rights* (Philadelphia: University of Pennsylvania Press, 2006).

documents of our human history, embodying the hopes and dreams of people still scarred from two world wars, newly fearful of the Cold War and just beginning the great liberation of peoples that came about with the dismantling of the European empires. It proclaims the fundamental freedoms of thought, opinion, expression, and belief and enshrines the core right of participatory and representative government. But just as firmly and with equal emphasis, it proclaims economic, social, and cultural rights and the right to equal opportunity. It was to be 'a common standard of achievement for all peoples and all nations', and the rights and freedoms it set forth were to be enjoyed by all without distinction. The fiftieth anniversary was an opportunity to remind the world that we needed these values as much in 1998 as the drafters did in 1948.

In my personal spiritual thinking, I had made a link between the concept of God being love and the first article of the Universal Declaration: 'All human beings are born free and equal in dignity and rights. They are endowed with reason and conscience and should act towards one another in a spirit of brotherhood.' Naturally I baulked at the reference to 'brotherhood' and not 'sisterhood', but understood the sentiment in the context of its time. The drafters, under the guidance of their chairperson, Eleanor Roosevelt, had drawn on the great religions of the world, and also on the humanist tradition, in defining the rights and freedoms they set out. More and more I was drawn towards the idea of a global ethic, where religious and faith leaders of the world, men and women, would come together and work towards a shared set of moral principles.

Above all, I felt we should understand the Declaration as a living document. I saw the anniversary year as an opportunity to advance the global debate and give more priority to current complex issues: the right to development out of grinding poverty, the recognition of the rights of indigenous peoples, the rights and empowerment of people with disabilities, gender mainstreaming, and issues of benchmarks and accountability in furtherance of these and other rights.

I was conscious of the need for a balance between the two streams of human rights, between civil and political rights – including freedom of expression; rights to be heard and to be free from torture;

and the economic, social, and cultural rights – to an adequate standard of living, including basic education and health care. I found that
developing – and particularly Asian – countries focused more on the
latter stream; they wanted to build up their economies and strengthen
health and education before opening up on civil and political rights.
Often this was the excuse used not to make reforms in dictatorial
regimes. Conversely, the more developed countries would emphasise civil and political liberties and often viewed the economic and
social rights as merely aspirational and somehow less important. The
challenge was to encourage countries in all regions to see the need
to advance the entire agenda and adjust their priorities accordingly.
It was a huge task.

The travel agenda that year was intensive. It came on top of the
work of building up the office, supporting independent experts, and
attending to the complex inter-governmental politics of the UN
Commission on Human Rights, strengthening the growing number
of field offices dotted around the globe, and reacting to violations of
human rights in real time. My days were hectic and gruelling, with
little time for reflection or rest.

I was missing Tessa and William, but got into the habit of phoning
them about once a week, when I could snatch a little bit of time, to
keep in touch. 'What airport are you in now, Mum?' they would
tease as they answered my call.

Because of the Universal Declaration's anniversary year, more
dignitaries, and at a higher level, were passing through Geneva to
talk about human rights. My philosophy was that every country has
its own human rights problems and that, as High Commissioner, I
should raise those issues and demand accountability. One of the
strengths of the office and of the broader UN system was that I was
able to receive detailed briefings about human rights issues anywhere
in the world. When a dignitary made a courtesy call to the office
to see me, he or she would often express concern about what was
happening in some other country. The German ambassador, say,
would bring up the conflict in the former Yugoslavia. After discussing these matters, I would then, very deliberately, turn the focus to
rights violations occurring in the dignitary's own country. In the

German example, I raised the problem of Turkish migrant workers. The dignitary would generally be quite surprised, and then try to defend his or her position or claim that the issue had been exaggerated.

US secretary of state Madeleine Albright came to the annual session of the Commission on Human Rights in Geneva that year, and her visit signalled recognition by the United States of the importance of the international human rights systems, which I welcomed. Madeleine Albright was taken aback, however, when in our bilateral meeting I raised my concerns about the US Government's own human rights shortcomings, particularly around use of the death penalty, and its not ratifying certain key covenants or conventions, such as those relating to economic, social, and cultural rights. 'A word in your ear, Mary,' she said. 'Don't worry about the United States. We can take care of ourselves.' But, as I replied to her, if I did not even-handedly raise my concerns with every country, if I let any country off the hook, I wouldn't be doing my job of demanding accountability without fear or favour.

Inevitably a new High Commissioner brings a change not just in personality and style but also in pace, vision, and priorities. In my first year on the job, I had managed to raise the office's public and UN profile, resulting in much greater scrutiny of our work, to say nothing of the dramatic increase in the volume of correspondence and requests for assistance and information. Our rising profile was not, however, matched by increases in personnel or finance, leading to a dangerous situation of unmet expectations on the part of those who wanted action combined with a small staff working longer hours, harder, and under greater pressure.

We were under constant scrutiny from those states that resented the role of the office, notably Algeria, Egypt, Pakistan, and Cuba. The diplomats from these countries seemed to wake up every morning thinking about how to keep the office pinned down, and I had to wake that bit earlier to try to stay ahead of them.

One of my greatest difficulties early in my term was having the flexibility to make my own decisions on hiring good people for key positions. UN bureaucracy threw up many hurdles on this path.

Understandably, the different regions of the world must be proportionately represented in any UN office. Unfortunately, this meant in some cases that the best person for the job would be disqualified because of where he or she came from. On top of the regional balance rules, all recruitment had to go through New York and could take many months – time I didn't feel we could afford to lose. This was like sand in the wheels of the machine.

Several times, in private meetings with Kofi Annan, I raised the need for increased staff, particularly in the liaison office based in New York. Kofi was sympathetic, but creating new posts in human rights is a difficult and sensitive issue.

A critical post was that of Deputy High Commissioner. Thomas Hammarberg, a former secretary-general of Amnesty International serving at the time as UN special representative in Cambodia, and a close friend and adviser, would have made an excellent deputy. But he was Swedish, a Northern European like me, and therefore ineligible to serve as my deputy. Enrique ter Horst of Venezuela was appointed deputy in February 1998, but found the job particularly difficult and had personal reasons to leave after only seven months. An experienced UN official, Bertie Ramcharan of Guyana, finally came on board to replace him in December 1998.* Bertie was an insider, had worked his way up through the ranks of the United Nations, had excellent drafting skills, and was known for his capacity to deliver, to get things done.

Gradually I did begin to find ways to get good people into key positions in the office. Stefanie Grant, a former senior staffer with Amnesty International and Human Rights First, would join us to lead the office's research division, strengthening our expertise on the links between human rights and development. Jan Cedergren, a Swedish Government official with a long track record in development work, came to lead our growing field operations. Brian Burdekin from Australia was already doing innovative work encouraging national human rights institutions. Ronan Murphy, an Irish

---

* Bertie Ramcharan went on to serve as Acting High Commissioner after the tragic death of my successor, Sergio Vieira de Mello, in Iraq in 2003.

diplomat with extensive experience in challenging posts around the world, including as ambassador in Moscow, took a leave of absence in 1999 to serve as my senior adviser and director of communications.

Our growing profile as an office was soon matched by new premises. During my first week as High Commissioner, I was invited by the Swiss ambassador to the United Nations, together with Nick, to visit a building the Swiss authorities were renovating and intended to turn over to the UN to house my office, on a magnificent site by the shore of Lac Leman: the Palais Wilson, so called because it had been the original headquarters of the League of Nations in whose creation US president Woodrow Wilson had played such a pivotal role. Climbing over rubble, watching the patient artisan work being done on an elaborate staircase, I wondered how long the building works would take and if the building would, as the ambassador assured me, be made available to house the Office of the UN High Commissioner for Human Rights. I could see the practical value and the symbolism of having our own building, but it seemed a faraway dream.

Then, just before 10 December 1998, fifty years from the day on which the Universal Declaration was adopted by the United Nations, the dream came true. Notwithstanding some political infighting by certain governments who resented the prominence these new premises gave to the office, and thereby the human rights cause, we moved into the Palais Wilson. The move marked another step in the maturing and strengthening of the office. It made a huge difference to efficiency and morale to be together in one location.

Bride Rosney's stint as my special adviser came to an end in the summer of 1998. I was deeply appreciative of all she had done to help me, but knew well how she loved Ireland and that she never wanted to live anywhere else. I would miss her humour, her direct, straightforward approach, and her ability to cut through UN bureaucracy, but I knew she would always be at the other end of a phone if I needed to talk.

We hired Jack Christofides, an urbane, accomplished South African diplomat with a strong commitment to human rights, to take over relations with delegations in Geneva and some of the media work that Bride had been handling. Jack was to have a baptism of fire on my first visit to China.

Early in 1998, China's permanent representative in Geneva, Ambassador Wu Jianmin, a strong, forceful character, attended my office personally with an invitation to visit China. I welcomed this as an indication that China, with its abysmal human rights reputation, wanted to improve its record and acknowledge a UN role in its affairs. But I signalled at that meeting that if I were to visit China, Tibet would have to be a part of the itinerary. For me, the ruthless oppression by the Chinese authorities of the Tibetan culture and religion was a serious blot on China's record, and one that required scrutiny. Ambassador Wu politely stated that visiting Tibet would be out of the question, and for several months we had a stand-off on the issue, but I nevertheless began to make the arrangements to travel to China in September 1998 – the first visit by a UN High Commissioner for Human Rights.

In August, Ambassador Wu came with detailed plans for the ten-day visit. I listened with increasing interest but had to put it to him that I hadn't committed because I had not yet been given permission to visit Tibet. He brushed this aside and continued to outline the itinerary. 'When you come to Shanghai,' he said, 'you will see a sign in the park dating from colonial times which we've kept. It reads, "No Dogs or Chinese" – we feel strongly about this.' I realised my moment had come, looked him in the eye, and said, 'Ambassador, I understand profoundly what you mean, because I am Irish. We, too, were colonised, we, too, suffered signs reading "No Irish Need Apply".' He looked at me intently and asked me to send him a history of my country. So I arranged for two or three Irish histories to be sent to Ambassador Wu, and a week later I received a message: 'Yes, you may visit Tibet.'

The challenge now was to raise the issue of human rights abuses in China without offending or alienating our hosts. To do this I would try to emphasise my political independence; that I was coming with a human rights rather than political agenda.

I had been keen that a number of international journalists would accompany me to Tibet to help convey to a global audience the significance of the visit. We had many requests through the Geneva office and directly from China. RTÉ, the Irish broadcaster, decided to send Charlie Bird, who duly received clearance. Then I got word that the *Irish Times* journalist based in China, Conor O'Clery, had obtained permission from the Chinese authorities to come to Tibet. I was glad to hear this, but concerned, as a UN official, that the only journalists being given permission were Irish. So I pushed back through Jack Christofides and sought to have permission for other international journalists as well. We still had no confirmation of this when we arrived in Beijing. I said – in retrospect most unwisely – that I didn't want to travel with just two Irish journalists, that I believed there should be a number of international journalists allowed. As a Western journalist based in China, Conor understand-ably felt he had earned his exclusive coverage and reacted sharply, as did the Chinese authorities, in a clever way, because they cancelled Conor's permit to come to Tibet, and I ended up with no print journalist. The situation had not been handled well, but then it was hard to organise a clever handling from afar. Beijing had taken advantage of what appeared to be an internal, Irish row to minimise my media coverage in Tibet. How I wished Bride Rosney had been with me! She would have judged the situation better and diffused it without controversy.

After the official visits in Beijing, we made the long journey to Tibet, stopping overnight. I learned from Charlie Bird that a huge row was breaking out at the *Irish Times*, the headline on the front page screaming, 'Robinson vetoes visit by reporter O'Clery . . .', and the *Irish Times* editor, Conor Brady, proposing to file a complaint against me. But despite all the fireworks at home the visit to Tibet, in our terms, turned out to be quite a success. The tools at my disposal as a UN official were limited: listening and respecting, bearing witness, providing advocacy, amplifying voices. Later I would bring any complaints and allegations to the attention of the government officials who had responsibility. I would also report on the visit to relevant UN bodies.

We spent three days in Lhasa, Tibet's capital. We were high up in the mountains, as high as I had been in my life. The air was clear and clean, but thin, though I was lucky to find that, unlike some of my colleagues, the altitude did not affect me.

The beauty of Tibet struck me immediately, with its high mountains and breathtaking views; but it was the spirituality of the people that made the biggest impression. You could see it everywhere, in the faces of Tibetans in the street, the strong presence of monks and nuns, their reverence in Buddhist temples. It was easy to distinguish the Han population, who had received incentives to move to Tibet to make it more 'Chinese'. Their presence reminded me of the plantations in Ireland of people from England and Scotland that I had learned of as a child.

Although we were shadowed at all times by officials, I felt the warmth of the welcome by ordinary Tibetans. The UN office in Beijing had provided us with an interpreter, and he tried to interpret as best he could the quick sentences of Tibetans, who knew they could suffer for speaking against the regime. 'A nun has been imprisoned and we hear she is on hunger strike.' 'We are not allowed to speak openly.' 'We want His Holiness here with us.' It was difficult for us to be discreet, as, understandably, Charlie Bird was keen to film us for an Irish audience far away. We were offered generous hospitality, the one problem being the offer of tea laced with butter that I was expected to drink on every occasion!

I visited the Potala, the palace in the centre of the city, once the seat of power of the exiled Dalai Lama. At a school in Llasa, I handed out copies of the Universal Declaration of Human Rights that had been translated into the Tibetan language, despite one of my more cautious officer's warning that this might be contrary to Chinese law. The visit gave me a real insight into the rich culture and the remarkable spirituality that form the sense of identity of the Tibetan people. That culture was under threat of being subsumed into the wider Han population's culture and consumerism. Their cause was truly a human rights one.

Back in Beijing, I met with Foreign Minister Tang Jiaxuan and raised with him my concerns about what I had witnessed in Tibet, and issues such as arbitrary detention, religious intolerance, and

China's notorious 're-education through labour' programme, and individual cases of dissidents being detained without charge. Everything, of course, had to be conveyed through an interpreter. This made communication more difficult and time-consuming, so that from the long list of issues I brought with me, having been briefed by my own office and by agencies such as Amnesty International and China Watch, I had to get straight to the core arguments without much possibility for elaboration. The body language of the men I met was often inscrutable, and their responses curt, so that I wondered sometimes if I was making any headway at all. When I raised the issue of 're-education through labour' for example, trying to explain how this practice – a system where those speaking out against the regime would be arbitrarily sent away to be 're-educated' for several years – was unacceptable, the minister responded rapidly in Chinese. I waited for the translation, which boiled down to, 'We don't have enough lawyers for due process.' When I asked how many lawyers there were in China, he replied, 'About a hundred thousand and very few in the rural areas.' This, for a population in 1998 of about a billion. 'Some Western countries,' I replied, 'think they have too many lawyers. I could arrange to send a boatload.' This made him smile.

I met with the Vice-Premier Qian Qichen, with whom I also raised Tibet in particular, and my visit culminated in a meeting with the president Jiang Zemin. With the foreign minister, I had signed a memorandum of intent on future human rights work, effectively opening an ongoing dialogue on human rights between the United Nations and the government. This soon led to China signing the International Covenant on Civil and Political Rights (having signed the International Covenant on Economic, Social and Cultural Rights shortly after I came into office) and agreeing to engage with my office in a series of rights-based workshops, the first of which would concern 're-education through labour'.

By signing this memorandum of intent, China was showing an initial willingness to address its serious human rights problems. I felt that at some level at least a psychological barrier had been broken. Part of the reasoning may well have been to avoid the political embarrassment of a resolution against China for its poor human rights record being adopted by the Human Rights Commission, but I also felt there was some openness to a deeper dialogue, provided it was respectful. I would seek to keep these lines of communication open, to progress things slowly and steadily, step by step, knowing that one could not change China overnight.

The workshops enabled my colleagues to bring international human rights experts to meet with their Chinese counterparts and discuss aspects of the rule of law; human rights and the police; the judiciary; and education about human rights.

In all, I paid seven visits to China during my five years as High Commissioner. The Chinese Government, particularly at that time, was providing huge opportunities for improved education and health care, and was proud of the fact. Later some health and education was privatised, and as China got wealthier, standards in these areas deteriorated. But at that time there was a big commitment to lifting people out of poverty and improving education and health. The Chinese mentality was much more open to what we term economic, social, and cultural rights than that of many countries in the West, and I gave credit for that to the Chinese.

In March 2001 China ratified the International Covenant on Economic, Social and Cultural Rights, with one reservation, rejecting the right to form free trade unions. I welcomed the ratification but regretted the reservation.

Because I was a former president, when I visited China for one of the workshops my office organised as part of our agreement with the government, I was afforded a meeting with President Jiang Zemin and with the vice premier, Qian Qichen. This was unusual for any UN official other than the secretary-general.

My meetings with Jiang Zemin were very formal, with chairs seated side by side and a government interpreter sitting behind each of us. Jiang Zemin spoke relatively good English and would greet me in English, but when we sat and conversed, he would speak only Chinese. We had a number of conversations, in the very first of which he raised the issue of Tibet. He had been to Tibet himself and claimed the Chinese were making great progress, that it had been very feudal and backward under the mediaeval 'splittest' Dalai Lama. I could only reply that I hoped the people of Tibet would be enabled to preserve their culture and identity, but I could see that we were miles apart in terms of our thinking.

I had already raised human rights concerns on individual cases earlier, with the vice foreign minister, Wang Guangya, my counterpart for this purpose. The conversations with Jiang Zemin were more philosophical; we talked less about individual cases and more about the human condition in general. At one stage, during a meeting in February 2001, we were speaking as usual through the interpreter, and I could tell by Jiang Zemin's tone that the question was of special interest to him. Then he broke into English and said, 'I'm particularly interested in your answer.' I waited for the translation. His question was: Given the incredible advances in scientific knowledge, did I as High Commissioner believe there was any role for religion in the modern world? I instinctively felt I should answer in a 'Chinese way', and start where he had started from. So I began by agreeing about advances in scientific knowledge – we'd just had the mapping of the human genome – and he was pleased with this part, nodding. 'But,' I said, 'if you're asking me as High Commissioner, I believe, for better or for worse, that religion continues to play at least as important a role in today's world as in the past and will continue to do so, as is evident in the Muslim world, and in those countries practising the major religions in every region.' He was not particularly pleased with my answer, and the meeting came to an end shortly after that.

My next meeting with Jiang Zemin was later that year, in November, after the terrorist attacks of 9/11. I recalled for him the question he had asked me at our previous meeting, about whether religion was still important in the world. He interrupted, saying those who had committed the attacks were not religious; they had been living in the Western world. I said, 'But you must realise that the driving force is based on a fundamentalist approach to religious conviction.' He seemed to be truly at a loss to understand fully what had motivated those who carried out the terrorist acts on 9/11.

During my visits to China, I was able to interact several times with two large civil society organisations, the China Disabled Persons' Federation and the All-China Women's Federation. Both were close to government, and yet appeared to be exercising some influence on the issues that concerned them. Deng Pufang, president of the China Disabled Persons' Federation, spoke openly and frankly, in very good English, about human rights issues. His words may have been influenced by his own background. The son of the Venerable Hai Deng, he had been under house arrest during the Maoist regime (for speaking out against it) and, I was told, had either jumped or been pushed from a high window, sustaining injuries that resulted in his being wheelchair-bound as a young man. When he invited me to visit the fine facility in Beijing for people with disabilities, he made it clear that this facility was an exception and that there were problems of discrimination and poverty for people with disabilities throughout China. He told me about the federation's attempts to take initiatives, and gave as an example a training hospital in Beijing, where visually-impaired people were being trained to massage and were thereby given an opportunity to work in productive jobs.

My most memorable meeting with the All-China Women's Federation was on my last visit to China as High Commissioner in August 2002. I was invited to travel with the president of the federation to a rural area about two hours' drive from Beijing. I discovered that, in China, *rural* meant 'only about 80,000 people'. The district was poor. We met representatives from the Women's Federation in a school that was quite run down, barely furnished, with paint peeling. We were there to witness how the local community was tackling domestic violence. As we sat around, the local committee explained to me how a legal office had been opened to advise victims, and the law was being promoted through graphic novels and other accessible literature. Then we were entertained by a local theatre group enacting four short dramatic scenes illustrating incidents of domestic violence. The first began with a woman whacking her daughter-in-law, and being outraged when the daughter-in-law produced a leaflet and said she was going to the legal office to complain. I had an excellent interpreter, but so vivid was the performance that it needed no translation.

Had my mandate been extended as I had hoped, after the attacks on the United States on 9/11, I would have liked to continue to build up links with both federations, which count many millions of Chinese in their membership. It is important and effective to encourage a culture of human rights in such organisations, and not just confine the dialogue to the difficult one of trying to hold the government to account.

On each visit to China, I had been more critical and outspoken on human rights concerns than any other UN official, but I had also acknowledged progress, such as the signing of the two international covenants and the subsequent ratification of one of them, and the work the office had been doing towards strengthening the understanding of the rights in question through the series of workshops. I continued to make it clear that my criticisms were on pure human rights grounds, without any political agenda.

At the end of each visit, I would give a press conference, always packed with Chinese and international press, that took place at the office of the resident coordinator of the UN in Beijing. At that last press conference, as I usually did, I gave credit to China not just for the useful meetings we had had but also for the series of workshops in which they had engaged. But I also expressed concern about ongoing human rights issues in the country: the imprisonment of political dissidents and labour protesters, further abuses in Tibet, and the growing number of cases of ill-treatment of individual Falun Gong members. I held that regardless of how the Chinese characterised Falun Gong, whether as a cult or otherwise, they mustn't abuse individual rights.

I was asked questions by the Western press – *The Wall Street Journal, New York Times, Washington Post, Financial Times* – and by Chinese print and broadcast media. I always tried to be straight in responding to questions, to answer the questions posed as best I could. At the end I said, 'I know what's going to happen after this press conference because it has happened after every single press conference I've held since I first visited China: the international press are going to cover what I said critically about China, and the Chinese press are going to focus on the positive things I've said.

Could we not for once have a balanced representation of what I'm saying?' There were a few titters of acknowledgment from the journalists. Sure enough though, the next day, the Western papers wrote, 'UN High Commissioner criticises China's record on . . .' and the Chinese papers proclaimed, 'High Commissioner praises China for . . .'

Plus ça change!

14

Bearing Witness

They are begging us, you see, in their wordless way,
To do something, to speak on their behalf
Or at least not to close the door again.

Derek Mahon, 'A Disused Shed in Co. Wexford'

THROUGHOUT MY time in office I had to think constantly about how a small and underfunded office could make an impact on advancing human rights and holding governments to account where there was wide-scale impunity. I realised that I could play a significant role by being close to the victims, bringing out their accounts, which was why so much of my time as High Commissioner was spent travelling to many of the most troubled regions of the planet, where people's rights were being violated and they longed for somebody to alert the world to what was happening to them.

In all, I made 115 trips to more than seventy countries during five years, almost always with the idea of helping to amplify the voices of victims, helping them to feel that somebody was listening. It brought home to me the power of the act of bearing witness. This was something I had encountered when I visited Somalia as president of Ireland. The act of witnessing is neither easy nor as forthcoming as might be expected. We turn away so often. I did not want to let that happen. Even though I held a UN title, I had nothing tangible to offer victims who were expressing their direct witness of torture, how their families had been killed, how they had been deprived of their land, their homes. They needed our action, not our tears; our practical, downright, problem-solving help, and not our wordless horror. Yet I felt that to listen, bear witness, and respect the humanity of those I was listening to and report back to a jaded world, was a start. I wanted to nurture a sense that the United Nations understood that these voices mattered.

There are some visits that stay in my mind because they illustrate perfectly the different facets of the work my colleagues and I were doing, and the kind of human rights violations we were dealing with.

In early 1999, world attention focused on the conflict in Kosovo, and the huge exodus of refugees driven out to different countries, especially Albania. In May the Swiss Government put at my disposal a small plane to visit countries of the region and draw attention to the plight of victims and to the human rights violations being perpetrated by Serbian forces. Over a few days we travelled to Albania, Croatia, Bosnia–Herzegovina and into Serbia itself by road from Zagreb. My two most vivid memories involve Kosovar refugees in Tirana, Albania, and ordinary Serbian civilians in Niš, who had just endured NATO cluster bombing of their housing estate.

Visiting the makeshift camp in Tirana, crowded with Kosovar refugees still traumatised after fleeing violence and savagery by Serbian forces in Kosovo, I listened to some of the worst stories while sitting in small tents that housed about fifteen people, seeing the fear still in the eyes of the children. My human rights officers were compiling initial evidence from witnesses that could be helpful to the International Criminal Tribunal for the Former Yugoslavia in the event that leaders such as Slobodan Milošević were indicted. At one stage a colleague suggested I come to a field behind the tents where a teacher was instructing a class. We watched as the children, sitting on the grass, repeated the teacher's words; they had no pens or paper. When she finished, we were introduced. She spoke good English, had been a teacher in Pristina, and had volunteered to teach the children in the camp. 'This is what I do; it helps me to cope.' She brought me to meet her elderly parents, sharing a tent with two other families, and described their escape from their home in Pristina, how they had watched it burn as they drove away.

The following day, my colleagues and I were in Belgrade meeting the foreign minister and other officials to raise serious concerns about human rights. President Milošević refused to meet me, perhaps because a few days before, I had met with the prosecutor of the

International Criminal Court, Louise Arbour, in Sarajevo. Milošević may have feared that Arbour had given me an indictment to serve on him. Indeed, within a matter of weeks, and despite political pressure from some sources, an indictment for his arrest was issued. Though, as we now know, tracking him down took years.

My colleagues were keen that I travel to the city of Niš to meet the mayor, who had, at great personal risk, been in contact with our office to report human rights violations. With a Serbian military escort, we set off by road, a journey of about two hours. We saw flashes of bombing in the distance. Shortly afterwards our small convoy was stopped and we were informed that NATO forces had just bombed part of Niš. I feared we would have to abandon the visit. We waited, and then suddenly we were rushed at some speed to a suburb, where we saw the immediate aftermath of the cluster bombs NATO planes had dropped. No one in that suburb had been killed, but several people were injured; a man invited me into his room in a high-rise block of flats that had suffered a direct hit and showed me a live, unexploded cluster bomb on the window ledge. As I came back down I was told the mayor had arrived, and we had a brief, tense meeting, where he quickly updated me and then left; he was clearly worried about being seen with me.

Our convoy returned to Belgrade before nightfall. There had been much emphasis by NATO forces on how targeted bombing was, hitting only military targets. I had seen for myself that this was not always the case.

Shortly afterwards I reported to the UN Human Rights Commission in Geneva on the violations in Kosovo I had witnessed and the plight of Kosovar refugees. I also took the opportunity to raise my concerns about the scale of the NATO bombing and the need for NATO's response to be proportionate. This caused a sharp intake of breath in the packed chamber. My words outraged, in particular, the United States ambassador to the United Nations, Richard Holbrooke, and I was subsequently reprimanded by Kofi Annan for criticising NATO at such a sensitive time.

What I was trying to convey – and this is one of the most difficult human rights challenges, something people don't usually want to hear, especially in times of military intervention – is that even when dealing with the worst, most egregious offenders, a minimum standard of human rights and international humanitarian law must be applied by civilised nations to protect civilians. That is the true test of legitimacy. The Geneva Conventions of 1949 and their additional protocols, particularly the Fourth Geneva Protocol, on the protection of civilian persons in time of war, made up the basic international humanitarian law binding on all parties. Civilians in housing estates must not be regarded as merely 'collateral damage'; they are men, women, and children who must be protected as far as possible.

I put on a brave face in the office but at home, with Nick, I let my guard down. The job was lonely and difficult, and the criticism of the way I was doing it hit home. Nick had the right response: 'Mary, today is Friday. Let me remind you there are such things as weekends. You and I are going to drive to Annecy to spoil ourselves.' Within minutes, packing for the weekend, my mood had lifted.

It was important to be strategic and balanced geographically on the issues of human rights violations we would seek to highlight. With so much political and media attention focused on Kosovo, my office began to receive complaints about the world's disproportionate attention to a human rights and humanitarian crisis in Europe. What about Sierra Leone, where there had been conflict for months and now fighting in the capital, Freetown, with many massacres? I took the point and was keen to respond. A number of African ambassadors invited me to lead a high-level delegation to visit the country and report back on the human rights situation there. I felt I needed an African champion to help me lead such a delegation and sought the advice of these ambassadors as to who would be appropriate. 'Try Ketumile Masire, former president of Botswana,' they advised.

I recall how hesitant I was when I finally got through on the phone to this former president I had never met. His response was immediate: of course he would do it. We were joined by the chairperson of the African group at the Commission on Human Rights, Ambassador Kamal Morjan of Tunisia; the Swedish human rights ambassador, Catherine von Heidenstam; and the executive director of the International Human Rights Law Group, Gay McDougall, and travelled together to Sierra Leone in June 1999.

Crossing by UN helicopter from Conakry, I was struck by the destruction in and around Freetown. I saw the roadblocks manned by Nigerian peacekeepers to prevent the rebels, who had been forced out, from returning. When we landed and began to meet survivors, I was shocked by the extent of the maiming – the hacking off of arms, hands, and legs by machete. I was haunted for months afterwards by images of what I had witnessed: an elderly man without hands learning to eat with two hooks; a pregnant woman, children pulling at her dress, with only one arm and one leg; three girls who told me how they had escaped the rebels after months of daily rape and slave labour; a camp for former child soldiers in Freetown, the children's only desire, to find their parents again.

Under the quiet, effective leadership of Masire, we met all parties, including representatives of the Sierra Leone political factions, the UN force, and the vibrant civil society. Together with President Kabbah of Sierra Leone; the secretary-general's special representative Mr Okilo; the head of the UN peacekeeping force; and the chair of the coalition of non-governmental organisations, I signed a human rights manifesto. This was an unusual initiative that helped forge a broad consensus on the need to protect human rights and bring to justice those responsible for violations, as began to happen subsequently with the establishment of the International Criminal Tribunal for Sierra Leone. Afterwards, I wrote to President Kabbah offering the support of human rights officers, and in particular Brian Burdekin, to help establish the truth and reconciliation commission referred to in the manifesto.

A few months later the attention of our office was focused on a small island far away from Sierra Leone. Serious violations of human rights had broken out in East Timor in August 1999, following the plebiscite for independence from Indonesia organised under UN auspices. I felt I should go there to highlight the plight of the victims, and to support Ian Martin, who was in charge of the UN presence in Dili. I flew to Darwin, in Northern Australia, where I met many

Timorese refugees who had fled the violence. Indonesian soldiers and militia forces loyal to Indonesia and opposed to East Timor's independence had kidnapped many Timorese and dragged them to their stronghold in West Timor. The violence had worsened while I travelled to Darwin, and I was told that the United Nations would not authorise me to cross over to Dili, as it was unsafe. This was confirmed by Ian Martin on the phone. His UN compound had in excess of a thousand refugees inside it, and more were climbing in each day.

Instead, having gathered first-hand accounts from a number of refugees and UN officials who had spent time in East Timor, I decided to fly to Jakarta and seek a meeting with President Habibie. Shortly afterwards Ian was told by UN security that he and his small team should evacuate for safety reasons. To his immense credit, he courageously refused, choosing to stay on in Dili and thereby protecting the lives of many. After the shameful withdrawal of the United Nations from Rwanda in April 1994, it would have been unconscionable for that to happen again.

Arriving in Jakarta, it was good to learn that President Habibie would meet me. He seemed taken aback and genuinely concerned when I gave him graphic accounts of what I had heard from East Timorese refugees while in Darwin, and that what was happening in East Timor – the killings and kidnappings – could amount to crimes against humanity and had to stop. The situation did ease a bit, but there's no way I can assess how much my contribution played a part in that.

During the visit, I learned that Xanana Gusmao, a former priest and hero figure of East Timor, was being sheltered in the British Embassy, and went to see him. He was an impressive man, quietly spoken and with a presence that conveyed a strong sense of moral authority. Speaking with the help of an interpreter, he told me that he didn't hate Indonesia. Far from it. He was determined to try to build reconciliation between his tiny country and its larger neighbour, in the long-term interests of his people.

In September 1999, I reported to the UN Security Council, the first time a High Commissioner had done so, and referred to the situation in East Timor.

As High Commissioner for Human Rights, I have assumed a burden of listening: listening to the pain and anguish of victims of violations; listening to the anxieties and fears of human rights defenders. I am glad to share this burden with you today, members of the Council, because you have the power and possibilities to alleviate the pain and to prevent some of the anxieties being realised. . . .

The awful abuses committed in East Timor have shocked the world – and rightly so since it would be hard to conceive of a more blatant assault on the rights of hundreds of thousands of innocent civilians. The murders, maimings, rapes and countless other atrocities committed by the militias with the involvement of elements of the security forces were especially repugnant because they came in the aftermath of the freely expressed wishes of the East Timorese people about their political future. I saw evidence of a well-planned and systematic policy of killings, displacement, destruction of property and intimidation. There must be accountability for the grave violations committed in East Timor. My recommendation is the establishment of an international commission of enquiry to gather and analyse evidence of the events in East Timor.

Back in Ireland the plight of the East Timorese seemed to touch a chord with people, and I read in the papers about the heroic efforts of a bus driver, Tom Hyland, who had collected money for humanitarian aid that he then brought to the country himself.

In August 2000, I finally made my first visit to East Timor. My colleagues and I had accommodation on a ship anchored off the coast, as it was deemed too dangerous for us to stay overnight on land. There were many signs of the trauma suffered by the people of this small island. I have a vivid memory of travelling from Dili to the village of Suai, a few hours away. As we arrived I heard women keening and saw them gathered around a shrine of stones and ribbons, marking the first anniversary of the massacre in the local church and the forcible taking of a number of men to West Timor; the men were still missing.

I sat and listened, through translation, to a number of widows and wives of missing men. A young woman offered me her small baby to hold. As I held the baby girl, the mother told me, 'Her name is Maria, like yours. She is my child by rape, and I am trying hard to love her'.*

A year later, in August 2001, I was honoured to address the national parliament of the Democratic Republic of East Timor, and to congratulate the country on the democratic institutions it had put in place. I encouraged the government to accede to the international human rights treaties, which it did shortly afterwards. On that visit, I also attended a hearing in a rural area before a regional commissioner of the newly established truth and reconciliation commission. I was pleased to see that the commissioner was a woman, wearing a sash to show her authority, sitting with the elders of the village. Three young men, referred to as the perpetrators, sat facing the community with their heads bowed. They had joined the Indonesia-supported militia mob that had burned houses in the village, but none of the three had killed anyone. At the request of the commissioner, I addressed the community briefly and commended them on this dignified and democratic process of reconciliation. The three young men, having expressed deep regret for their actions, were allowed to return to live in the village.

This was one of a number of opportunities I had to sit in on truth and reconciliation commissions in different post-conflict countries. My first experience had been in South Africa in 1996, during my state visit there as president of Ireland. I met Alex Boraine, who was the deputy chair of the South African Truth and Reconciliation Commission, serving under Archbishop Tutu. Alex Boraine went on to found the International Center for Transitional Justice in New York, to support a growing number of truth commissions in places as far apart as Peru, Burundi, Sierra Leone, and of course, East Timor. 'Transitional justice' is the term used to describe all the measures taken by different countries to redress the legacies of massive human rights abuses, including criminal prosecutions, truth commissions and

---

* Several years later I received a note from the same mother with a photograph of a happy young child beside her. Somehow the mother had come through her trauma; her maternal love had won out.

various types of reparations programmes and memorials. Because of the numbers involved, criminal prosecutions can deal with only a small fraction of cases. While there are concerns about impunity, the truth commission model can be effective in this type of situation as it enables many more victims and families of victims to tell their stories, confront perpetrators, and occasionally receive some form of reparation.

One of the most difficult but significant missions I was ever involved with came about in the spring of 2000. Early on in the annual six-week session of the Commission on Human Rights, in March, I began to receive credible information about human rights atrocities being committed by Russian soldiers and government forces in Chechnya.

The second Chechen war had begun in late August 1999, when Russian forces responded to the Chechen invasion of Dagestan, mounting a series of massive air strikes over Chechnya followed by a ground troop invasion to reoccupy the region.

The prevailing view in Geneva was that no High Commissioner could get away with taking on Russia, one of the five permanent members of the UN Security Council. But Ronan Murphy, who had been the Irish ambassador in Moscow from 1995 to 1999, had come on board as an adviser and was instrumental in negotiations with the Russians to obtain permission for me to visit the Chechen region.

The Russians were unenthusiastic to say the least. They denied any human rights abuses and were patronising in the tone they took, claiming that we didn't understand what they were dealing with. In the battle of wills, I think I must have been the more stubborn, as eventually, reluctantly, they agreed to let us in to Chechnya and to supply transport – provided, however, that we also visited Dagestan, to see the impact of the Chechen incursion there.

Unexpectedly, then, for a High Commissioner, I left the Commission on Human Rights mid-session and set off with Ronan Murphy and our human rights desk officer on the region, Tanya Smith. Our visit began in Nazran, the capital of Ingushetia, a region bordering Chechnya and by then hosting tens of thousands of people who had fled the fighting. We visited the main camp, where thousands of people were living in tents. I also walked through some of the ninety-six railway carriages, home to about four thousand displaced persons. One carriage housed forty-five people from sixteen different families. They had been there for six months. Word of our visit spread, and I was approached by many people, mainly women, who were distressed and made appeals for help. They were deeply concerned about what the future held for them, about missing relatives – particularly those left behind in towns and villages in Chechnya that had been bombarded, and those taken to detention centres – about lack of health care, restrictions on travel, and difficulties with identity documents. Tanya and a local human rights advocate translated for me and took notes of what was said.

We were then brought to an overcrowded tented area. There were so many people there pushing in, eager to speak to us, that it did not always feel safe. Our Russian escort stood aside and made no effort to offer me protection. It occurred to me that, in the murky way of these things, they would have shed no tears if any injury had befallen one of us.

The intrepid Russian human rights organisation Memorial had arranged that later that evening I would meet a number of victims in a house nearby, where my colleagues and I could ask questions and take notes for our report. I was deeply impressed by the courage shown by Memorial in helping to highlight complaints levelled against its own country in a conflict situation. For several hours I heard detailed first-hand testimony from witnesses of alleged gross violations of human rights by Russian military, militia, and Ministry of Interior forces in Chechnya: I heard allegations of mass killings, summary executions, rape, torture, and pillage. I was given photographs and videotape evidence and shown the wounds and scars of those who had themselves been injured. The individuals were clearly traumatised by what they had endured, and frightened, but they gave detailed, precise answers to close questioning – they were eager to be accurate.

I listened to the account of a woman who, with two other women, had gone back to the Staropromoslovsky district of Grozny on 21

January to check on their houses. She described abuse at check-points, insults, ransom extorted, and finally how the three women were blindfolded by troops whom she described as being from the regular army. The women were then taken to a destroyed house and the blindfolds were removed. The women pleaded desperately, but the witness described how the soldiers shot the first woman and how part of her head came off; how they then shot her, but the bullet went through her shoulder and she collapsed. While she was semi-conscious, her earrings and ring were pulled off and her clothes searched for money. The next sensation was of burning on her leg when mattresses were placed over the three bodies and set alight. The witness managed to crawl away and was brought to a cellar where other people were hiding. Eventually she was able to be reunited with her family. Her injuries were serious and required urgent attention.

The next day we flew by helicopter to Chechnya's capital, Grozny, accompanied by a Russian general straight out of central casting: burly, stern-faced, in full uniform topped with the distinctive over-sized military cap. We were taken around Grozny by coach, the general and other soldiers with us. The whole city had been demolished, both by the latest offensive and during the first Chechen war. And yet it was quite extraordinary; out from under the rubble would emerge an elderly man or a woman with a child, coming to a cross-roads where there were small stalls with food, or a distribution point. The coach stopped at each of these crossroads so we could get out and try to get as much information as possible from the remaining residents of Grozny, these survivors.

At about the third stop, the Russian general went to the door of the coach and started yelling in Russian at the gathering crowd. I asked Ronan what he was saying. It turned out he was telling those outside, 'Stop complaining! There are to be no complaints at this stop. You are forbidden to complain.' It was unbelievable! But of course they did complain, and bitterly, of lack of sufficient food and of their miserable living conditions. Many were anxious about relatives who had been detained. Criticisms were made of the Chechen fighters for their callous disregard for the welfare of the civilian population. A simple point made by many Chechen women was 'We are not all bandits.'

After this, we were required, as a condition of the Russians providing our transport to Grozny, to go to Dagestan. We were to take a helicopter on an hour-long flight to Dagestan's capital, Makhachkala. We got into a primitive helicopter, with wooden seats at the sides, no seatbelts, no ear protection against the noise. In the middle sat a tank with fuel; we placed our luggage around it on the floor. The general didn't come with us, but we were accompanied by two heavily armed soldiers. After about an hour my colleague Tanya Smith pointed down, and I glimpsed a city beneath the clouds. She gave a thumbs-up sign. But we seemed to be circling, and didn't land immediately. I subsequently understood that the president of Dagestan wanted to be at the airport to greet me (coldly) and that he wasn't ready when we were overhead, so our pilot had been ordered to circle. But as we circled, the weather completely closed in and the helicopter clearly lost its way. For another hour we circled, and circled. At this stage one of the soldiers took out a pack of cigarettes and proceeded to smoke one after the other, nervously, right beside the fuel tank. I thought to myself, *This will be a very strange end.* Tanya caught my eye as if to ask, 'Are you going to say anything?', but I wasn't going to tell this soldier to put out his cigarette. It was very frightening. It was frightening, too, when we were coming down; clearly we were making something of an emergency landing: we broke through the cloud, and not far below us was a field. We'd missed the airport by a long way.

A man driving by in a tractor and trailer was quickly required to take one of the soldiers to some place where he could make a phone call. We stood there waiting in the dark for a long time before cars eventually came to collect us and drive us, perhaps for an hour, to the capital of Dagestan; we had flown that far off course.

By now it was about ten o'clock at night. We were scheduled to have had a dinner with the president at eight o'clock, so I assumed that had been cancelled and we could just go to bed. But no, we were to have the dinner. When we arrived at the hotel we went upstairs briefly to freshen up and then entered a room where guests had assembled, presumably having been there for hours. Two of them, women, were poets, and we fell into an easy conversation. Then I took my seat at the table – this was now about midnight – opposite the empty chair of the president. The president came in, stood at his place, and proceeded to make a long and hostile speech, lecturing me about the mistake I had made in being sympathetic to the Chechens because, after all, it was they who had invaded Dagestan, and so on. Then he sat down and indicated that I could now speak.

I stood up. I remember wondering how to play it. First, I thanked him for the hospitality; the table was laden with food. Then I said that I was sorry that because of the circumstances, though not through any fault of my own, it had not been possible for us to meet earlier. I could sense the president squirming slightly. I thanked him for bringing together such distinguished guests; I'd had an opportunity to meet two distinguished poets, and I smiled at them and they smiled at me. The president was getting a little confused now, because I was saying nice things. I acknowledged the trauma the Chechen attacks had had on the Dagestan people and then said that, as a human rights person, I sought to be fair to every country I visited and to understand the culture of that country. I told him of the plight of the refugees I had met in Ingushetia, and the many allegations of gross violations that I had heard in Grozny. Perhaps he could use his influence, I suggested, as a friend of the Russian authorities? Then, finally, we ate. At the end, when the president said good night to me, after a long meal and endless toasts, he shook my hand and said, 'You are not so bad!'

We flew from the capital of Dagestan to Moscow the following day. Ronan was able to tell me that in some of the newspapers, I was being described as 'the great white witch'. As we boarded the old Russian plane, with windows so dirty you could not see out them, Ronan assured me it was about as safe as the helicopter we had travelled in the day before!

When we met with the foreign minister, Igor Ivanov, in Moscow – President Putin had refused a meeting – he tried very hard to exclude our interpreter and just have the one, his. But I insisted that our colleague be present. To control the interpreter is to control the conversation. The foreign minister spoke at some length, and eventually I interrupted; it was a real tussle. If we hadn't had our interpreter in the room I couldn't have proceeded, because I couldn't trust his to convey what I was saying to him. With my interpreter standing behind me, I looked the foreign minister in the eye and described, frankly and thoroughly, all that we had heard and seen.

Feeling somewhat relieved that we had survived the trip intact, we then flew straight to Geneva, where the Commission on Human Rights was still in session. Despite the opposition of the Russian authorities at every pass, the gruelling conditions experienced during the visit, the terrible stories we heard, which had the absolute ring of truth, we managed to bring back an account of human rights abuses on a pervasive scale committed by a permanent member of the Security Council. Ronan and Tanya had worked flat out on compiling my report, and I added my own impressions and sense of urgency. I delivered the report to a packed, hushed Commission, and was followed by the permanent representative of the Russian Federation, who gave an angry riposte denying any human rights violations. Nevertheless, and for the first time in its history, the Commission on Human Rights adopted a resolution censuring a 'permanent five' country (meaning one of the five on the Security Council who have a veto), a measure of how affected the delegates had been by the account of what was happening to the people of Chechnya. A week or so later, Ronan proudly showed me an article on the visit to Chechnya in the *New York Review of Books*, which drew extensively on my report to the Commission.

Standing up to the Russians over Chechnya had been very satisfying, even though, as is the way of these things, no individuals were actually brought to justice as a direct result of my report. When I eventually arrived home to Nick in Geneva I was mentally and physically exhausted but Nick could tell from one look at me that I was also elated. 'Well, you do choose your battles!' he said.

Inevitably, the Israeli-Palestinian situation was an important focus in our work. The second 'intifada', as it was seen by Palestinians, began in September/October of 2000. A special session of the Commission on Human Rights concluded with a resolution requesting that the High Commissioner and various rapporteurs visit the region to examine the human rights situation and report back to the Commission. I asked Kofi Annan to seek permission for me to go. The mission, to Israel and the Occupied Palestinian Territories, and also Egypt and Jordan, was conducted from 8 to 16 November 2000. My mandate for the mission meant particularly focusing on the situation in the Occupied Territories; it meant addressing issues of excessive use of force, the impact on civilians, (particularly children), discrimination, inequality, humiliation, and the very serious economic situation. But I did not want to get drawn into the political dimensions of the ongoing conflict.

It was my first time in Israel and the Palestinian Territories, and the first thing that struck me was how small the areas are and how close together everything is. I was shocked by the impact of the conflict in the Occupied Territories. The high number of killed and seriously injured was having a devastating effect on the civilian population. There was a destructive effect on children, generally, in their opportunity for education, due to so many schools being closed and, as teachers emphasised, how difficult it was for young people to concentrate in such situations. In Gaza, I was told that it was very difficult for patients – ill people and pregnant mothers – to go to the West Bank or Israel for medical treatment.* There were complaints about degrading treatment at checkpoints, although there was, at least, movement of people. But the reality of occupation was one of daily grinding humiliation that must over time build up a sense of utter frustration and anger. This had come to the boil in the recent intifada, and was exacerbated by expansions taking place in the areas of settlement, something I found almost incomprehensible when I saw evidence of it; the settlements were causing such an amount of tension.

In Israel the main complaints were of 'drive-by shootings' of settlers on the settler roads. I condemned these in my report as being completely unacceptable, as was any violence against civilians. These

---

* But they could go; it became much more difficult later.

settler families were living in fear, and there was a great feeling of insecurity and fear in Israel itself. At their request, I met some of these families; indeed I met with a wide cross-section of authorities and people both in the Occupied Territories and in Israel. I was also able to raise key issues with senior officials of the Israel Defense Forces and meet Israeli human rights groups. The Israeli Information Center for Human Rights in the Occupied Territories, known as B'Tselem, was particularly impressive because it was taking cases in the Israeli courts on behalf of Palestinians who wouldn't be allowed to plead on their own behalf: a humbling example of human rights courage.

The most striking visit was to the city of Hebron, a place of long-standing tension, which had flared up during the intifada. Hebron had both a Palestinian and an Israeli section. In the Israeli-controlled areas, there was a relatively small number of Israeli settlers, several hundred maybe, and several thousand Palestinians, who, because of the rising tension, were kept under a strict curfew that amounted to virtual house arrest. They were allowed out only at certain times to shop, and were otherwise ordered to stay off the streets. As a result, it was quite eerie driving around these empty streets.

Our small team of guides in Hebron, young Scandinavians, were members of the TIPH (Temporary International Presence in the City of Hebron), an international civilian observer mission stationed in Hebron. Our convoy was just two vehicles, a minivan, which carried most of us, and a car in front. Our guides required us to wear flak jackets; it wasn't entirely safe. At one checkpoint, several settlers, who were on their way to collect their children from school, banged on both vehicles in anger as we passed; they obviously knew who we were and didn't like our being there.

We then drove up a hill so that our guides could show us Israeli installations – listening devices – that had been put on top of a couple of Palestinian houses and a school. While they were showing us this, I heard the crack of gunshot: two shots were fired very close to us, one of which lodged in the car. We all dived back into the minivan and crouched down on the floor. Both the car and the minivan sped over the hill and stopped in a lane. We got out, rather shaken. The shots had been close; I had been conscious of a bullet passing.

The young Scandinavian team rang their contact in the Israel Defense Forces to arrange protection for us. While they were doing this I took out my phone and rang my father. I didn't tell him where I was or what was up; we simply chatted, about nothing at all really. I just felt that if anything did happen he needed to have heard my voice one last time. I admit I was frightened.

We were instructed by the Israelis on the phone to get back into our minivan and car and drive down that hill again, the way we had come, because it was the only road out; they would be at the bottom of the hill waiting for us. We were to keep our heads down. So we crouched in the back of the minivan while the driver took us down, and when we got to the bottom, a number of Israeli soldiers were indeed waiting. I raised an eyebrow as they told us that it was probably Palestinians who had shot at us. Essentially, the Palestinians wanted us there, the settlers didn't. So why would the Palestinians have shot at us?

There was quite a bit of media interest in this incident, but I was anxious to play it down. I didn't want it to prejudice the report I would be submitting; I was absolutely determined to be fair, balanced, and objective. In my report, I spoke of the completely different narratives on both sides. I did feel shocked at the separation, at the settler roads that the Palestinians weren't allowed on, at the fact that settler houses were still being built in both the West Bank and Gaza at the time.

The report was well received by the Commission in Geneva. I remember being praised for balance by the then-US ambassador Nancy Rubin. However, because the report referred to the root of the problems stemming from the occupation of Palestinian Territories, the Palestinians praised the report widely, with the result that Israel claimed that it lacked balance. It was a lesson to me in just how difficult it was, despite the best intentions, to strike a balance that was perceived to be fair by both sides: virtually impossible.

My lasting impression of the region is of two narratives of victimhood: The Palestinians feel the victimhood of the occupation and their second-class status, the curfews, the Israeli settlers getting all the best land and continuing to build houses. The Israelis feel victimhood because they are being attacked, missiles are being fired from

Gaza and killing innocent civilians, and because the United Nations has had endless resolutions criticising Israel; they are constantly in the dock. My mandate required much of my report to be taken up with the human rights situation in the Occupied Palestinian Territories, but I believe Israel had a fair point when it argued that no other country had come under the same level of UN scrutiny. Israel has independent judges, courts, a democratic electoral system; many other countries with much worse human rights violations never face a permanent item against them on the UN agenda.

The result: both sides felt they were victims and both sides felt that the United Nations was inadequate. And, in effect, the solutions from each side's point of view seemed utterly incompatible.

Although missions I undertook to countries in conflict, or with long-running situations of human rights violations, could be difficult – and at times dangerous – I felt this was an essential part of my job. It gave me a better understanding of the role of UN special rapporteurs, who had mandates to focus on particular countries, such as Iran or Somalia, or on theme areas, such as extra-judicial killings, torture, or violence against women. I admired the courage and tenacity of these frontline human rights defenders, men and women of strong human rights commitment and expertise, who worked without pay and often travelled in difficult circumstances to report on violations and to hold governments accountable under the covenants and conventions they had ratified.

My own direct experience of witnessing violations of human rights and listening to the accounts of victims has left me with a series of vivid images just below the surface of my active consciousness, images that are easily recalled if I discuss a situation or some passing reference acts as a trigger. At one level, it is a burden to have these troubling memories seared in my mind. At another, it has given me an enduring empathy with courageous victims who are often agents of change in their very tough situations.

15

Into the Crucible

It is neither easy nor agreeable to dredge this abyss of viciousness, and yet I think it must be done, because what could be perpetrated yesterday could be attempted again tomorrow, could overwhelm us and our children. One is tempted to turn away with a grimace and close one's mind: this is a temptation one must resist.

Primo Levi, *The Drowned and the Saved*

IN 1997, with the Rwandan genocide of 1994 still fresh in the collective memory, the United Nations had sought once more to address racism and xenophobia. A resolution was passed to hold the third World Conference against Racism, Racial Discrimination, Xenophobia and Related Intolerance in the autumn of 2001. The General Assembly's decision to hold the conference came during my first year as High Commissioner, and my office was assigned the task of organising the conference, which would be attended by delegations from most countries, hundreds of NGOs, labour organisations, UN officials, and others. I was both daunted and excited by the challenge.

Preparation for the Durban conference started in earnest about eighteen months beforehand, and required an enormous effort on top of the heavy workload of the office. I was painfully aware that the United Nations had organised similar conferences on racism, in 1978 and 1983, both of which ended in shambles, with no consensus reached. I was determined to shepherd governments toward a meeting of minds on the sensitive topics to be discussed – topics so crucial to human rights.

At this time, in the run-up to the conference, I was considering whether to seek a second four-year term as High Commissioner, which would begin immediately after the conference closed, on 11 September

2001, or whether to regard what I hoped would be the successful outcome of that conference as my last contribution to the office.

In the end I decided not to seek a second term, for several reasons, foremost among them being that I lacked a clear signal from Kofi Annan about his support. While attending a meeting of the UN senior management team of under-secretaries-general in New York in late 2000, I received a neutral response when I asked him about my future as High Commissioner. He suggested that it was a topic for further discussion. But he did not follow up with me, and I was left wondering whether I would really have his backing should I decide to go forward.

Unquestionably, I had been an outspoken High Commissioner. I knew that Kofi, whose role as Secretary-General necessitated conciliation and diplomatic compromise, was on occasion less than comfortable with my assertive style. But I took seriously the UN Charter, which begins, 'We the peoples of the United Nations . . .' That's who I believed I represented, the people – not the governments – of the member states. That was how I kept faith with my job.

I was also weighing up the burdens of the office. The job had been gruelling, the most difficult of my life, and by early 2001, perhaps not surprisingly, I was getting pressure from Nick and our children not to seek a second term. Nick put it in positive terms: I had transformed the office, and now was the time to hand over to someone else.

As a courtesy, I let member states know I would not be seeking a second term during the 2001 annual meeting of the Commission on Human Rights in Geneva. I made my announcement during my opening speech on the first day the Commission was in session, 19 March 2001. 'This has not been an easy decision to make,' I said, 'and I know it is one that will surprise and perhaps disappoint many, [but] I believe that I can, at this stage, achieve more outside of the constraints that a multilateral organisation inevitably imposes.' I promised to support the United Nations 'in whatever way I can.' I truly felt confident that I was leaving the office on a stronger footing, with improved morale and a higher profile and influence within the United Nations and throughout the world.

I was unprepared for and ultimately very moved by the response my announcement engendered. I was inundated with messages from people all over the world saying, essentially, 'Please, stay!' I found it particularly poignant that so many of these pleas came from the grass roots.

I was lobbied by heads of state, too. Among the most insistent of them, intriguingly, were Jacques Chirac of France and the new president of Algeria, Abdelaziz Bouteflika. I had visited Algeria twice, and perhaps Bouteflika believed that I was a non-Muslim who was at least trying to understand his country's difficulties. Chirac was even more insistent, and I believe his support had an impact on Kofi Annan, when he and I eventually had a meeting about my future at the United Nations.

I confess I was at first quite emotional when I met Kofi at his hotel suite in Geneva during the Commission meeting, to discuss the possibility of my staying on. I explained to him that part of my reason for wanting to leave had been what I perceived as his ambivalence concerning my second term. He was warm to me and explained that he had been preoccupied with various problems and had not intended to leave me guessing or to make me feel I lacked his support. He said, 'You did a good job. I would be happy if you would stay on, and I know there's a great groundswell of support for you to stay.'

Nick had been delighted that I was not seeking a second term, but now he could see how torn I was. 'What about one more year?' I suggested. 'Could we manage that?' Reluctantly, Nick agreed, and I proposed this compromise to Kofi. I hoped that this additional time would help ensure that whatever consensus document emerged from the Durban conference would have practical consequences. The follow-up period, I believed, was going to be as important as bringing the conference to a successful conclusion. Kofi and I drew up an agreement during a meeting in Nairobi following the session of the Commission on Human Rights and he brought this proposal before the UN General Assembly. By July 2001 the arrangement had been formalised – I would remain in office until 11 September 2002. I was relieved and energised by the response I had received.

Speaking at the launch of a business plan for the Pavee Laundry at Belcamp Lane, Dublin, April 1995. It was the first such business to be set up by Traveller women.

A well-wisher tests my good-natured 'ring of steel' on a Dublin walkabout.

Laying a wreath in September 1992 at the war memorial in Enniskillen, scene of a bombing outrage.

With Queen Elizabeth II at Buckingham Palace, May 1993: the first official meeting between Irish and British heads of state since independence in 1922.

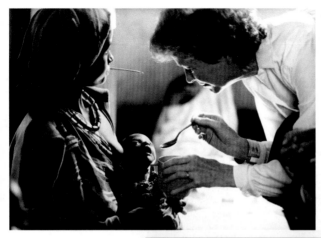

Following my visit in October 1992 to famine-stricken Somalia I told the international press of the horrors I had witnessed: '... I cannot be entirely calm speaking to you, because I have such a sense of what the world must take responsibility for'.

On his state visit to Ireland in 1996 President Vaclav Havel chose a refreshingly informal venue to host his return dinner, Johnnie Fox's Pub in the Dublin Mountains.

A happy visit organised by the Northern Ireland Council for Ethnic Minorities, Belfast.

In the White House during a state visit in June 1996, with President Bill Clinton and Vice President Al Gore.

Meeting Pope John Paul II in March 1997. It was International Women's Day, hence the sprig of mimosa pinned to my coat.

In front of the newly renovated Palais Wilson in Geneva, which was handed over to the United Nations by the Swiss government to house the office of the High Commissioner for Human Rights, June 1998.

With Secretary-General Kofi Annan at UN headquarters.

September 1998: one of several
interesting meetings with President Jiang
Zemin during my seven visits to China
as UN High Commissioner.

With Nkosane Dlamini Zuma, president of
the 2001 Durban World Conference Against
Racism, held just a few days before 9/11.

An emotional moment as President Barack
Obama confers the US Presidential Medal
of Freedom, July 2009.

As a member of the Mo Ibrahim
Foundation I enjoy Mo's refreshing
directness, good humour, and occasional
political incorrectness. October 2008.

With two of my heroes, Stéphane Hessel and President Jimmy Carter, promoting the Elders' campaign 'Every Human has Rights' on the 60th anniversary of the Universal Declaration of Human Rights, December 2008.

April 2011: landing in North Korea on an Elders mission to discuss dangerous food shortages and stalled nuclear disarmament talks, with (from left) Martti Ahtisaari, Gro Brundtland and Jimmy Carter. Behind him is CEO Mabel van Oranje.

Realizing Rights in practice: with Nyaradzayi Gumbonzvanda, General Secretary of the World YWCA, bringing a group of women leaders to Zimbabwe to show solidarity with women working to foster democracy and human rights, April 2010.

The Dalai Lama enthrals a huge audience at the Town Hall, New York, May 2009.

Archbishop Desmond Tutu is another spiritual leader who skilfully uses humour to engage his listeners. Here he teases me during an Oxfam International climate hearing in Copenhagen, December 2009.

Nelson Mandela with the Elders (from left) Graça Machel, Fernando Henrique Cardoso, Desmond Tutu, Jimmy Carter, Mary Robinson, Kofi Annan, Gro Harlem Brundtland, Martti Ahtisaari, Ela Bhatt, Lakhdar Brahimi.

And yet, as fate would have it, my final year as High Commissioner would be a year like no other, a year of startling international upheaval, with new, insidious threats to human rights everywhere.

Every nation, however large or small, has something to hide about racism and xenophobia: take Canada with its Inuit people, the Czech Republic with its Romani population, France with its migrants, Ireland with its travellers. The base morality of colonialism has been a powerful negative influence, but the caste systems in many parts of the developing world, as well as institutional racism, even now, in this twenty-first century, are undeniable factors.

Also to be considered are the movements of Asian and Caribbean populations into the United Kingdom; North Africans from Algeria, Morocco and Tunisia into France and now into slums outside Paris; the large Turkish population in Germany. These migration issues would also be on our agenda.

When I was studying at Harvard in 1968, a number of my contemporaries travelled at great personal risk to join the civil rights movement seeking to address blatant racism in the southern states of the United States. While the country has made great strides and most persistent forms of racial hatred may have receded, more subtle forms of racism continue for African Americans, Muslims, and other minority groups.

Ambitiously, the Durban conference sought to bring together a vast range of people with a stake in eliminating this horrific global scourge. The enormous cast included delegations from most UN member states, civil society from every part of the world, and the various UN agencies for which these issues are relevant.

Alongside the UN conference, a conference of NGOs and civil society participants, including business, would also be held in Durban. In total, eighteen thousand people, not including the media, would be in attendance. All the groups would be vying for influence over the wording of the final consensus declaration and the global programme of action.

Playing host, as a nation, to a World Conference against Racism, Racial Discrimination, and Xenophobia and Related Intolerance is not exactly like holding the Olympics. Apart from South Africa, just one other country, Brazil, volunteered to be host, and even Brazil's offer was half-hearted. By 1999, the United Nations deemed South Africa to be the better choice, and accepted its offer. In my office we were relieved and delighted that the conference would be held in Durban. South Africa – a country once torn apart by Apartheid – had shown that racism could be defeated, that progress could be made.

As host country, South Africa would have the primary political leadership role at the conference, and thus Foreign Minister Nkosazana Dlamini-Zuma (a formidable woman leader, once married to the current president of South Africa, Jacob Zuma) would be president and chair. My role on the ground was to be somewhat limited: I would head the secretariat that would provide the services – documentation, translation, and so on – required by the diplomats and other participants in the conference. Member states would steer and negotiate the documents of the conference – a role they jealously guarded – and ultimate leadership would fall upon Zuma's capable shoulders. As the preparatory and conference process played out, circumstances determined that I had to play a proactive role as we fought to keep the project on course. Nevertheless, as a UN official, I was determined throughout the conference to keep Foreign Minister Zuma front and centre.

Delegates from countries in each of the UN regional groups gathered throughout late 2000 and early 2001 for preparatory-conference meetings, 'prepcoms'. The four regional preparatory events – held in Strasbourg, Santiago, Dakar, and Tehran for Europe, the Americas, Africa, and Asia respectively – were each charged with putting forward their region's collective views of what the eventual conference consensus declaration should contain. These regional prepcom documents were far from set in stone; their very fluidity was the point. They would be synthesised later into a final document in Durban.

It had been clear to me early on that holding one of the four regional conferences, the Asian prepcom, in Tehran, would be problematic. Iran, chosen by the countries of the region, was a poor choice to host a conference addressing issues of racism, xenophobia, and anti-Semitism because of its known hostility towards Israel. Tactically, it would have been better to move the preparatory meeting else-where in the region, but no other governments offered, probably because many had their own minority, caste, or racism issues.

In its Prep Comm session, the Tehran meeting, held in February 2001, harshly criticised Israel for its policies in the Palestinian territories and its treatment of Palestinians and made an analogy between those policies and Apartheid.* The 'Zionist movement . . . is based on race superiority,' the draft declaration subsequently alleged, along with the charge that Israel had carried out 'ethnic cleansing of the Arab population of historic Palestine'.† All such sentences were opposed by some delegates present and, as is always the UN procedure, were put in square brackets in the text, indicating they had not been agreed upon.

At the time, I felt certain that this inflammatory language would be removed from further draft texts well before Durban. Unfortunately, as the preparatory processes went on, the states that had inserted the bracketed language in Tehran refused to withdraw it.

Looking back, I realise I put too much store in the fact that any controversial clauses put in square brackets would either be removed during the preparatory process, or inevitably would be thoroughly debated during the tough negotiations on a final text. I underestimated the hurt and anxiety words in a document would cause, regardless of whether they were in brackets or not, and that the political fallout would start before the Durban conference itself.

In my capacity as secretary-general of the conference, I was not at the negotiating table itself and could only seek to lead states in the right direction, but nothing I could do would force them to follow. I felt one of the avenues open to me was to change the narrative on tackling racism, xenophobia, anti-Semitism, and other forms of intolerance; to grab the moral high ground with as many key leaders as I could, thus making it impossible for offending bracketed language to survive the process. One step I took was to create a vision statement that heads of state and government would be

---

* UN DOC A/CONF, 189/PC, 2/9, paras 20–21.
† UN DOC A/CONF, 189/PC, 3/7, paras 29 and 63.

invited to sign before their delegates came to Durban. The statement read, in part, 'We all constitute one human family . . . Instead of allowing diversity of race and culture to become a limiting factor in human exchange and development, we must refocus our understanding, discern in such diversity the potential for mutual enrichment, and realize that it is the interchange between great traditions of human spirituality that offers the best prospect for the persistence of the human spirit itself.'

Nelson Mandela 'godfathered' the statement; his signature was at the bottom, next to mine. I hoped Kofi would sign the statement as well, but the response I received from his office was that his signature 'wouldn't be appropriate'. It was a signal that the Secretary-General's advisers intended to distance him from what was clearly going to be a difficult conference.

I wasn't entirely surprised. With the vision statement, I had gone against orthodoxy and created an unusual, 'outside-of-the-box' initiative, and the United Nations doesn't necessarily like people at my level doing anything unusual. Nevertheless, in the months before the Durban conference began, my colleagues and I got signatures from more than eighty heads of state and government, and the vision statement was later referred to in the preamble to the Durban Declaration and Programme of Action.

The early months of 2001 were difficult for me for a very personal reason. My father was dying of cancer. As a doctor, he knew the cancer was well advanced, so he decided not to seek chemotherapy but to enjoy the quality of his last months of life. This he did in an extraordinary way, one that enhanced the lives of his wide circle of family and friends. Every week he would be visited by cousins, nephews, and nieces from far-flung places, and friends who would sit with him and share stories. We, his immediate family, spent as much time as possible with him, although, sadly for me, I was least able for obvious reasons.

Finally, in early July, at age eighty-seven, my father sent a note to his last remaining private patients saying he could no longer look after them. Days later he was moved to Castlebar Hospital, where he was tenderly cared for by doctors and nurses alike, some of whom he had helped bring into the world.

I joined my brothers and sisters-in-law and our young at his bedside, all of us coming in and out and sometimes all crammed together in the small hospital room. As a deeply religious man, he had no fear of death and was ready to meet his maker. He had positive words for each of us, his sense of wonder enduring to the end.

His death was peaceful in the true sense, he left behind him no unfinished business, and his funeral gave the people of Ballina an opportunity to demonstrate the huge affection and esteem they held for him. For me, the most moving moment was when some of his grandchildren began to recite prayers for the living and dead that they themselves had chosen.

Going back to work was a kind of therapy. My father's death changed me. Now I was a member of the older generation; I would have to try to be there for the younger ones in the way he had always been there so deeply for each of us.

In the final years of the Clinton administration, the White House had demonstrated its support for the proceedings in Durban. Bill Clinton appointed a group of high-level officials at the State and Justice departments to a task force on the subject, and the administration held a series of town-hall meetings on race around the country. But by February 2001 it was evident that the newly installed administration of George W. Bush had much less enthusiasm for the conference, and particularly after the Tehran meeting, their lukewarm interest turned to red-hot aversion.

I met with Colin Powell, the new US secretary of state, for the first time on 8 February 2001. Initially, I felt I had his unqualified support. Indeed, I knew he would have loved to come to Durban. 'But,' he told me, 'there are difficulties.' Of course, we both knew that. In spite of the problems, however, Powell continued to talk to me over the coming months, his goodwill always evident.

When the criticisms of Israel expressed in the Tehran draft became public that spring, members of the Jewish-American lobby began an intensive campaign in the American media demanding that the

United States withdraw from the conference. The threat of a possible walkout by the US delegation rapidly became a central theme of the Bush administration's negotiating strategy.

It was soon apparent that the Bush administration had another quarrel with the Durban conference: calls for a discussion of the thorny issue of reparations for slavery and a formal apology for slavery and the slave trade, raised by the African prepcom in Dakar, were also going to be on the agenda.

I met with Powell again on 18 June in Washington. He outlined the Bush administration's hard-line stance: if the bracketed language from Tehran were not removed, the United States would not participate in the conference. He also told me that although the United States was willing to express regret for its involvement in the slave trade, it was not prepared to use the word '*apology*'. Such an admission might have implications in litigation for reparations in national courts.

Immediately after my meeting with Powell, his spokesperson at the State Department, Richard Boucher, announced publicly Powell's threat to boycott the conference, saying I had failed to address US concerns. On 30 July, Congress passed Congressman Tom Lantos's sponsored resolution affirming support for the Bush administration's hard-line position on the Durban conference.

It was genuinely difficult for me to address the criticisms of Congressman Tom Lantos. I had admired him as a Holocaust survivor and a wonderful human-rights advocate in the US Congress. When I visited the United States as Irish president, he was my biggest champion and supporter in the Congress, and he wholeheartedly welcomed my appointment as UN High Commissioner for Human Rights. In that capacity, I visited Capitol Hill at his warm invitation. I very much regret that Tom and I never reconciled our differences about Durban before his untimely death in 2008.

It is understandable – but regrettable – that Tom Lantos and others did not fully comprehend the bureaucratic and formalistic ways of the United Nations when negotiating the documents of a world conference. In particular, critics failed to appreciate that wording put forward by some governments and noted in square brackets when consensus was not reached, was text that had no validity unless subsequently agreed to and taken out of those square brackets in final conference negotiations. The anti-Semitic language inserted by a small number of governments in square brackets at a regional conference in Tehran was duly removed during the negotiations in Durban and did not appear in the final text, but unfortunately, for many, the damage done by that draft language to the legitimacy of the final outcome could not be repaired.

At the end of July, I called another meeting in Geneva, this one unscheduled, to attempt once again to work out the conflicts over the bracketed texts. Lantos came to my office and was clearly furious that I wasn't able to get the offensive language removed. '*You* are the one who is responsible for removing this language,' he insisted. 'It's *your* conference. *You* are in charge of it.' He simply couldn't understand that I was, in this instance, powerless.

At a press conference on his ranch in Crawford, Texas, five days before the conference was to begin in Durban, President Bush announced his decision. Responding to a reporter's query, he said, 'We will not have a representative there as long as they pick on Israel. We will not participate in a conference that tries to isolate and denigrate Israel.' Shortly afterwards, I learned that Colin Powell would not be attending the conference but that, in his place, the United States would send a 'working-level' delegation.

Meanwhile, during the last week in August, I was already in Durban with most of the colleagues from my office who were part of the World Conference team. It was a small but dedicated team, led by my senior adviser, Ronan Murphy, acting as coordinator at a working level; Jyoti Singh, an experienced UN administrator whom we had brought in to manage the overall events and logistics of the Durban conference; Andrew Goledzinowski on communications; and Laurie Wiseberg, acting as liaison to the hundreds of NGOs that had come for their pre-summit conference. My deputy, Bertie Ramcharan, would follow the political discussions in detail, and stand in for me on the official platform when I was giving a keynote address at one of the important side events that were held on the margins of the conference.

The NGOs' conference took place in a separate venue nearby. There were vibrant meetings and events, but also the beginnings of a street sentiment against Israel, equating the treatment of Palestinians with the treatment suffered under the South African Apartheid regime. With sinking heart, I was realising that South Africa was a more problematic setting for the world conference than I had anticipated.

Two of my children – Tessa, who was by now practising law in Dublin, and Aubrey, who was studying at Cape Town University – joined me in Durban. They both took part in a rally organised by the South African trade union movement and some civil society groups, and returned quite shocked at the anti-Israel slogans and anti-Semitic sentiments of the large crowd. The participants were mainly South Africans, but included civil society representatives from around the world. The rally was an early harbinger of a mood that was to become even darker and more strident as the inter-governmental conference began.

On 29 August, Dr Dlamini-Zuma and I participated in a tree-planting ceremony in front of the Durban International Convention Centre to mark the event that would begin in earnest the following morning. Zuma, wearing a dazzling white jacket and floor-length skirt, addressed a gathering of UN staff. The trees we were planting were yellow-wood trees, the national tree of South Africa, so I wore a yellow suit. Zuma remarked that ten years earlier she would not have believed South Africa could play host to such a momentous conference.

I was seeking to build a relationship with Zuma that day, keen that she exercise full political leadership as chair and that we work well together. I was *so* determined that this conference succeed that I didn't even countenance the idea of failure.

'Madame President of the conference, President Mbeki, heads of state and government, Secretary General, excellencies, ladies and gentlemen, friends . . .' I began my address to the delegates of the UN member states on the first morning of the conference, 30 August. I spoke from a podium in the main auditorium of the Durban International Convention Centre, immediately following Kofi Annan.

I spoke slowly, conscious of the interpreters who were translating my words into six UN languages. I reminded the delegations from around the world that the success of the conference would be measured by the degree to which it brought solutions and relief to victims of racism. Fidel Castro and Yasir Arafat as well as the president of South Africa, Thabo Mbeki; the president of Latvia, Vaira Vike-Freiberg; Jesse Jackson; and Harry Belafonte were among the many political leaders, activists, and celebrities seated in the plenary hall that morning. In all, sixteen heads of state, fifty-eight foreign ministers and forty ministers had come to Durban.

My spirits were high. Success – meaning a Declaration and Programme of Action reached with consensus and with no offending language – seemed very much within grasp. As it turned out, success was to be forged in a trial by fire, and the conflagration began almost immediately.

The US delegation was led by Michael Southwick, a former ambassador to Uganda, who headed a small team of State Department lawyers and negotiators. Worryingly, the Americans arrived with an ultimatum that was contrary not only to the procedural agenda but also to the spirit of a world conference, the very purpose of which was discussion and negotiation. Unless the bracketed language that had been inserted following the Tehran regional conference was removed within twenty-four hours, Zuma and I learned to our dismay, the US delegation would decamp. Israel took the same position.

Zuma and others at the conference had repeatedly assured me that the offending language would come out by the end of the conference, but they insisted that negotiations must come first. Thus assured, I met with the Americans two days later, on Sunday 2 September. (At that point they had extended their ultimatum to thirty-six hours.) I told the US delegates that their heavy tactics were not helping and that their threatened refusal to participate fully in the week-long negotiations was 'warped, strange and undemocratic'. Later that day I met with Lantos and pleaded with him to persuade the American delegation to consider that we were just at the starting point of the discussions that would eliminate the offensive language. Lantos, in turn, told me the United States would hold me responsible for the conference's failure.

Early the following afternoon, on 3 September, President Bush and Colin Powell pulled the US delegation out of Durban. Powell asked Lantos to break the news to the international media gathered at the convention centre – more than 1,200 journalists were present in Durban for the event – and Lantos obliged, telling reporters that the United States had 'gone the extra mile' and had made an 'enormous compromise' by exhibiting a willingness to discuss the Middle East at the conference. By contrast, Lantos continued, Islamic countries had been rigid and uncompromising and had 'hijacked' the proceedings.

Gay McDougall, a distinguished international human rights lawyer and participant in the world conference as the first US member of the UN treaty body on racism, noted that the US walkout had been threatened for so long that when it came, it was anti-climactic. In her analysis of the conference, she noted, 'The US actions . . . seemed more like an objection to even having the debate, rather than being outcome-oriented. The US walkout was an abdication of global leadership and a shortsighted snub of the multilateral process.'*

Anti-climactic though the walkout may have been, its aftermath took on a crisis quality. Kofi Annan, who had already returned from Durban to UN headquarters in New York, immediately called me for an assessment of the situation. Shortly afterwards he issued a brief statement urging delegates to 'stay the course . . . The conference cannot afford other defections,' he concluded.

Later that day I was participating in a panel discussion with business leaders on fighting discrimination in the workplace when someone quietly handed me a note that said the conference was about to collapse as a result of the United States' and Israel's decision to leave. Rumours had begun circulating that Canada and members of the European Union would also leave. I left the room and literally

---

* Gay McDougall, 'The World Conference Against Racism: Through a Wider Lens' (2002) *Fletcher Forum of World Affairs* 26(2): 135, 143.

ran to the Hilton hotel to find Zuma in her suite, together with Belgian foreign minister Louis Michel, who was attending the conference as president of the Council of the European Union. They were silent, absolutely down. Michel had his hand on the phone. He was about to ring his colleagues to say, 'I am leaving tonight – Europe is withdrawing.'

I recall saying to Zuma, 'Either this entire event fails now, or you, – as chair – do what you can do to lift this conference up and take out the language in brackets. Do this tonight, and tomorrow morning we can go ahead. We don't need the Americans – South Africa can do this.'

Louis Michel had strong feelings about the damaging role played by Belgium in colonial Africa, and I knew he considered the conference to be an important step in addressing the strained relationship between Europe and Africa. I implored him to stay. I watched him put his pipe in his mouth and take a puff. 'Mary, you do not have to beg me,' he finally responded. 'I think this is an important conference, and I am here for as long as it takes.'

The problem was not yet solved. Michel still had to persuade the other European Union delegates, and Zuma had to take the reins on the matter of the anti-Semitic language.

Late that evening, just as I returned to my hotel room in need of a night's sleep, I received a phone call from Louis Michel, asking me to come to his suite. Ronan Murphy joined me, and he later described the scene we encountered there as if it were something Stanley Kubrick might have filmed. We opened the door onto a room filled by a massive square table at which fifteen heads of the European Union delegation were seated, each of them waiting to hear my arguments for why they should remain in South Africa. I sat down. I explained that I believed the South African president of the conference, Zuma, would expunge the bracketed language. 'What guarantees have you?' they asked. All I could say in response was 'I have no guarantees.'

At various times during the next several hours the Europeans asked me to leave the room, then invited me to return, as they pondered their course of action. I was persistent. I repeatedly called upon what by then felt like my mantra: the conference was a chance for important new language acknowledging the evils of racism and xenophobia and a prescription for how we might tackle these pernicious issues in the years ahead.

When I wasn't with the European delegates, I was in Zuma's suite. Zuma was in and out of meetings with South African president Mbeki, who was closeted in a back room with delegates. The two were making a feverish effort to expunge the long-reviled language on Israel and were stressing the importance to South Africa – and to Africa as a whole – that the conference succeed. I managed to sneak in a few hours of sleep, waking in time for a dawn meeting with the conference's steering committee, a balanced group whose members hailed from every region. Zuma announced that the language would be cut. A lengthy session ensued until the committee approved the new text with none of the bracketed language. When the meeting ended, it was apparent to all that we needed to go immediately to the full conference, where member states and civil society groups were waiting, and assure them we were moving forward to deal with all the other critical issues on the conference agenda.

As Zuma and I walked toward the plenary hall, she suddenly turned to ask if I would explain to the plenary meeting what had happened during the night. I told her she must do it, reminding her once again that she was the chair of the conference. She was profoundly exhausted walking into that room, yet she not only found her words, but also thoroughly re-energised the conference. 'Nothing is beyond discussion,' she told delegates. 'That is the beginning of a tolerant society – one that can sit down and negotiate, one that can listen to another point of view, even if you don't agree with it.' Europe decided to stay, as did Canada.

After the drama of the previous evening and early morning, there was to be another twist in the saga. I had scheduled a press conference late that morning, so I went to the South African broadcasting studio to get my make-up done. One of my advisers, Andrew Goledzinowski, found me in the make-up room. Andrew, phone in hand, told me that the US delegation leader, Michael Southwick, was on the line. Southwick was at the airport. He told me that, although he had instructions to return to Washington and was about to board a plane, the United States had *not* withdrawn from the conference. 'We will be there at a very low level,' he said, 'but we have not withdrawn.' The Consul General in South Africa would continue to represent the United States.

I told him, 'Can you go through that with me again – I want to be very clear – I have a press conference in a few minutes.' And so he repeated himself. I asked if he was aware that the language that caused his departure had been removed that morning. 'Yes, I know that,' he told me, adding, 'Why couldn't it have happened earlier?' I tempered my frustration – I was blue in the face at this stage from explaining why.

When it became apparent that the Israeli delegates were not aware of this development, I decided that Israel should be informed that the US delegation 'had not officially withdrawn'. In the meantime, I went to the press conference and faithfully reported, as I had been told by Southwick, that the United States was still in the conference. Two hours later my words were contradicted: the White House announced that both Israel and the United States had withdrawn from Durban entirely.

I never understood the full background, but thought it likely that when Andrew informed the Israeli delegation, contact was made directly by them with the White House and the US withdrawal was confirmed.

There were yet more battles to be fought in Durban. We were forced to extend the conference by a day to get the final text approved. Bertie Ramcharan was grim-faced with anxiety that consensus would not be reached. I had to plead with the interpreters to stay on through Saturday, 8 September. Some of the detail of those final hours is a blur. Kofi Annan was calling from New York; Javier Solana, the High Representative for Foreign Affairs and Security Policy of the European Union, was calling from Brussels. I was assuring them there would be a consensus final outcome. The UN media spokesperson for the conference had left on Friday, a rather graphic indication of how most of the UN system was distancing itself from anticipated failure.

Finally we were down to the last areas of debate, and I was able to persuade a Brazilian delegate to move for a procedural vote so that the final text, somewhat cobbled together, could be adopted by acclamation. And that was how it ended.

During the final press conference, Zuma made a metaphorical connection between the experience of the week just past and the pain of childbirth and subsequent joy of a new baby. Unlike the two previous conferences, and in spite of the pressures and walkouts and threatened walkouts, we had managed to stay the course. That evening, despite being sleep deprived and emotionally exhausted, I remember feeling an extraordinary, adrenaline-like high.

Although it was not a binding document in the full legal sense of being able to be relied on in a court of law, the declaration contained a solemn commitment by all the governments present. It meant that civil society organisations in each country, UN special rapporteurs, UN treaty bodies, and the UN Commission on Human Rights itself would be able to hold these governments to account for the commitments made.

Governments had expressed in the Durban declaration 'profound regret' to those who suffered the effects of slavery, the slave trade, the transatlantic slave trade, Apartheid, colonialism and genocide. Although it fell short of the apology many had sought, this was the strongest language yet to be used on those issues in an international document.

The process itself had brought governments to note 'that xenophobia against non-nationals, particularly migrants, refugees and asylum-seekers, constitutes one of the main sources of contemporary racism . . .'. The national plans of action and programmes that governments committed to on the broad agenda of the conference included tougher anti-discrimination legislation and better remedies for victims.

But the conference outcome has always been marred by the focus on the Israeli-Palestinian issue. While I have been praised in some parts of the world for the positive outcomes of Durban, I have also been reviled by some parts of the Jewish community, particularly in the United States, for what is viewed as a lack of effective leadership in Durban or even, for some, anti-Semitism. It means a great deal to me that this criticism does not come from the courageous Israeli human rights groups – B'Tselem and others and Israeli women's groups – who sprang to my defence when I was criticised by some Jewish groups as being unworthy of receiving the US Presidential Medal of Freedom from President Obama in 2009.

I have learned to be philosophical about those who express hatred and contempt for me as a result of my actions at the Durban conference or in other statements I've made on the conflict in the Middle East. It gives me a personal insight, which otherwise I might not have had, into the far greater pain of the slights and discrimination so many suffer because of racism.

16

When the Dust Settled

I came back from that frugal republic
with my two arms the one length, the customs woman
having insisted my allowance was myself.

The old man rose and gazed into my face
and said that was official recognition
that I was now a dual citizen.

He therefore desired me when I got home
to consider myself a representative
and to speak on their behalf in my own tongue.

Their embassies, he said, were everywhere
but operated independently
and no ambassador would ever be relieved.

            Seamus Heaney, 'From the Republic of Conscience'

THE DAY after the close of the Durban conference, a Sunday, Ronan Murphy and I took a long drive up the South African coastline. The beauty of the landscape was extraordinary and was balm for our physical and mental exhaustion. We didn't say much; we stopped at a fish restaurant and stared at the sea. On Sunday night I flew back to Geneva for a day's work; in the evening I went on to Dublin and then to my home in the West of Ireland.

The following day, 11 September 2001, is seared in my memory, as it is for so many. The day was gloriously sunny and my brother Adrian had promised Nick and me a treat: we would take a day trip by motorboat to the tiny Inishkea islands off the coast of Belmullet, County Mayo.

These two islands had a terrible sadness to their history. In the 1920s a storm killed twenty-seven fishermen, after which no one wanted to live on the islands. The families, comprising about 130 people, moved to the mainland, visiting the islands only to care for sheep and tend their land. The church, school, and houses fell into ruin on both islands, leaving a mysterious landscape for the visitor.

Our group that day included my brother Adrian; his wife, Ruth; the local parish priest and good friend, Kevin Hegarty; and our boatman, Vincent Sweeney. We anchored on the south island and had a picnic of smoked salmon, wine, beer, and sandwiches. We sat in the sunshine listening to the tale of how the islanders were politically divided and even threw stones from one island to another. Then we took the short boat trip to the north island to view an early stone carving depicting the resurrection of Christ, which Kevin wanted us to see. The ancient stonework, which dated from the eighth century, was heavily weathered.

As we were reflecting on the history of the place Vincent heard his radio phone bleeping and went up the hill to get a better signal. Seconds later he came running down towards us, his face white. 'I hate to tell you this, Commissioner, but a plane has hit the Twin Towers and another is heading for the White House!' The pallor of his face shocked me almost as much as his words. Silently, we returned to the boat. My brother and sister-in-law were sick with anxiety; their son and his fiancée were working as lawyers in firms close to the World Trade Center. Looking at my brother's haggard face, I took his hand and said, 'Adrian, whatever happens, I will always remember this as a day in two parts. Thank you so much for a glorious first part.' He was too numb to say anything.

For several hours we watched the images again and again, of the planes striking the Twin Towers on the small black-and-white television set in Vincent's home. We learned that my nephew and his fiancée were safe, but we could see that thousands had been killed, and horrifically.

When the dust had cleared from 9/11, literally and figuratively, I recall thinking to myself, '*What will this mean for human rights?*' That may sound a bit abstract, or theoretical, but in fact I had an ominous premonition that what would follow would undermine many core standards and values, painstakingly built up over many years and crucial to the protection of human rights worldwide. It dawned on me that, along with many fresh projects, ideas, hopes, the Durban Declaration and Programme of Action had been buried in the rubble of the Twin Towers. Few in the political world would have the patience for a rational analysis of all that had been accomplished in Durban and the steps taken there toward eliminating racism and xenophobia. In a matter of hours, the globe was turned on its axis; and the Durban conference, with all its controversy, was understandably out of the news and forgotten by the media – although, as things turned out, not forgotten by the individuals who had contested it so powerfully and then left before its conclusion.

I cut short my break in Ireland and hurried to New York. Over the next several days, in addition to myriad meetings at the United Nations I visited Ground Zero, spoke with family members of victims, with fire-fighters who had lost colleagues, with a vast array of volunteers, and with employees of FEMA, the disaster response agency, who had come from all over the United States to help. People from all backgrounds showed great spirit, helping and comforting one another, providing practical support of food and drink for those working and counselling for those who needed it. It was inspiring.

I remember one evening, leaving the pier where the Red Cross and others were assisting family members, and hailing two nearby taxis: one to take my colleague, Elsa Stamatopoulou, to the Tribeca neighbourhood where she lived, and the other to take me and my colleague to our Midtown hotel. By coincidence, both the drivers were Sikhs. They told us there would be no charge for the trips – they, too, wanted to show their support for all those who had lost their lives and all who were working tirelessly in the recovery efforts.

We held meetings in our UN office with human rights experts, debating how to respond to these terrible attacks. We concluded that the attacks of 9/11 constituted crimes against humanity, because the attackers had attempted to kill as many civilians as possible in a deliberate and premeditated way. I was vocal about the need to bring the perpetrators of these crimes to justice: terrorist attacks against innocent civilians are terrible violations of human rights, and it was important that all countries join in countering them, and bringing Osama Bin Laden and his associates (who were admitting their responsibility) to justice. But I continued to worry about the possible implications for the rule of law and the standards of due process.

My worries were confirmed by President George W. Bush's declaration of a newly conceived 'War on Terror', a war he promised would be waged by the United States and other Western nations. The hawkish tone of Bush's rhetoric was problematic; the rapid undermining of civil rights that ensued was devastating. In one of its first moves after the president's initial statements the US State Department rounded up more than a thousand people with Middle Eastern backgrounds, denying them their right to due process – on the strength of the unilateral declaration of war.

I saw then the myriad ways in which, under the rubric of national security and the imperative of protecting innocent citizens from additional terrorist attacks, this new war paradigm would potentially allow not only the United States but other governments to shirk obligations to protect human rights around the world. As I began my final year as High Commissioner for Human Rights, I needed to address something entirely new: a post-9/11 world. The concerted follow up on Durban I had been planning would unavoidably have to take a back-seat.

Following the meetings with UN experts, my office issued an early statement making the case that the attacks on the United States were crimes against humanity, a concept well-defined in international jurisprudence. Precedents and procedures are in place to deal with the perpetrators of such crimes. Hijacking planes full of people, full of jet fuel, and aiming them at buildings in order to kill as many civilians as possible was, by any standard, a crime against humanity. I tried to persuade the Bush administration that the way forward was to go after the perpetrators of these crimes and that police forces around the world stood ready to help bring them to justice. But such arguments seemed to have no effect on President George W. Bush, urged on by Vice President Dick Cheney and Secretary of Defense

Donald Rumsfeld, who opted for the language and strategy of a war against terrorism.

History has shown us that in war, despite the evolution of international humanitarian norms, there is still a sense that everything is fair game. Post-9/11, basic due process rights were an early casualty. In the United States, the Patriot Act, granting unprecedented new surveillance and detention powers to law enforcement and intelligence agencies, was enacted without any real debate in Congress. Detention camps in Guantánamo Bay were opened without regard for the Geneva Conventions. An attitude to torture, ambivalent or worse, surfaced. 'Extraordinary rendition' to torture-practising countries was carried out. Racial profiling was used. The United Kingdom enacted emergency laws eroding fair procedures. Worryingly, other countries with less convincing human rights records – Egypt and Pakistan, for example – opportunistically followed suit, further undermining the rights and freedoms of their citizens and of the press by the introduction of even more stringent emergency laws. A new language emerged; even political opponents could be characterised as terrorists.

My job became immeasurably more difficult. I became the United Nations' 'uncomfortable voice' on 9/11, insisting on speaking candidly and bluntly around these erosions in human rights standards. In mid-January 2002, I issued a release from the OHCHR in which I reminded the United States that suspected al-Qaeda members, by then detained at Guantánamo, were entitled to the protection of international human rights and humanitarian law. Any trials had to be guided by the centuries-old presumption of innocence. Complaints were made to Kofi Annan about my criticism of US policy. A significant sector felt that it was inappropriate to criticise the United States in any way, given the terrible events that had occurred. Journalists asked me why I would jeopardise my UN job by speaking out so forcefully. I replied that this *was* the job; it was better to *do* the job than try to keep it. But it was a lonely position; few official voices were publicly beating the same track.

At the same time, direct accounts were flooding into my office in Geneva from special rapporteurs with mandates on torture, extrajudicial killings and freedom of religion, warning of the deteriorating situations they were witnessing. Human rights organisations such as Amnesty International and Human Rights Watch were gathering evidence of abuses used in countering terrorism in various countries. When I tried to hold those governments to account, challenging, for example, a clampdown on political opponents or freedom of the press, their answer to me would be 'Everything has changed now; look at the United States.' No matter how hard I tried to uphold the rules, widespread slippage in the application of international human rights standards was evident in countries around the world.

Despite the urgency of the work that was arising in the post-9/11 context, there were of course innumerable other human rights issues that needed attention. As UN High Commissioner for Human Rights, I had become aware that many of the most serious violations of human rights are perpetrated against women. Women are constrained by fundamentalist laws, they are subjected to early child marriage, to genital cutting and other harmful traditional practices, and to distorted interpretations of religions that limit women's rights. Women and girls are subjected to slavery and trafficking and, in conflict situations, to large-scale rape and sexual violence at the hands of their oppressors. I also became aware of the huge numbers of brave women who work as human rights defenders around the globe, often at great personal risk.

Nothing brought this home more vividly than my visit as High Commissioner, to Afghanistan in March 2002, to coincide with another International Women's Day. I was welcomed by Dr Sima Samar, Minister for Women's Affairs under the provisional administration led by Chairman Karzai. Kabul was not completely safe at the time; the Intercontinental, where our delegation stayed, was really the only functioning hotel. A group of about ninety leading Afghan women – women who had been professors, lawyers, community leaders in the pre-Taliban era, some of whom had provided secret education for girls during that terrible period of repression of women – had been working, with support from Dr Noeleen Heyzer, executive director of UNIFEM, and her colleagues, to prepare for International Women's Day, drawing up a charter of women's rights.

Sima Samar explained to me that the venue for the event would be a cinema that local women had been able to attend in the old days but that had been destroyed by the Taliban. Local women had cleared out tonnes of rubble and the UN had provided a large tent cover. There was some concern that the women would be too afraid to attend, given the ongoing security situation.

On the morning of the event the venue was heavily guarded, and women who would ordinarily have worn the burka appeared without it. When I went to take my place on the platform, I noted with some irony that the first ten rows of seating had already been filled by bearded men. Behind them, the women filed in and took their seats at the back.

There were two other women on the platform with me, Dr Samar and Dr Heyzer. Chairman Karzai and Lakhdar Brahimi spoke, after which I also spoke, and congratulated the women on the work they had done in preparing a charter of women's rights.

Finally, after a long string of further speeches by men, a spokesperson for the women sitting towards the back of the audience stood up and talked about the agenda they were seeking to advance. Holding up the charter, she launched it to applause from her colleagues and some of the men. In the wider context of the uncertainty for these Afghan women as their country suffered further turmoil in the wake of the invasion, the gesture might have seemed hopeless. But these women were carrying out crucial work in the wake of the toppling of the Taliban. It was the first time since the attacks on the Twin Towers that I felt hopeful.

As the months unfolded, I began to discuss with Nick the idea of staying on in the job for a full second term. Though he understood how I felt, how it did not seem right, somehow, to depart at such a difficult time, he was not keen. He was looking forward to a time when I would not be working such relentless, draining hours but would be a private citizen again, able to relax more, spend more time with him and our family. But I felt compelled to stay on as High Commissioner, and Nick supported that decision. One of Nick's great qualities is that he is prepared to give way on his own strong preferences when he knows how deeply I feel about something. So he accepted the prospect of living in Geneva for another three years, which would mean postponing his own plans. We both worried that they would be lonely years, as Aubrey was by now studying in Cape Town.

On the margin of a senior management meeting of the United Nations in November 2001, I made my moment with Kofi Annan and said that I was prepared, if he would put it before the General Assembly for approval, to stay on for the three remaining years of this second term as High Commissioner. Kofi said that he would seriously consider this, and to leave it with him. He was aware, he said, of the American criticism of me, but that the United States was, after all, only one country.

Kofi Annan never came back to me on my proposal. I didn't hear anything, and the weeks went by. By early 2002 it was clear to me that I would not be asked to serve out a full second term.

On a personal level, I was upset and saddened that Kofi had not backed me, after all the work I had done to improve the office and raise its profile, and the risks I had taken in speaking out even-handedly on highly sensitive issues. Of course, I was very conscious of the pressures he was under. The United States Government made it clear that it was against my reappointment. But, perhaps naïvely, I thought this was all the more reason for extending my term, as I would not flinch from speaking out against any policy, Western or otherwise, wherever I took the view that it was contrary to the international human rights agenda. Another factor that inevitably influenced the Bush administration's views was its misperception of my role in the Durban world conference.

A bitter aftertaste persisted for the constituencies who had been so inflamed by Durban. As time passed, the Durban conference and the 9/11 attacks became somehow conflated in some people's minds. Disturbingly, Congressman Tom Lantos wrote, 'For us [who were] at the conference, it is evident that the same attitude that sought to turn Durban into an anti-Israel carnival also led to the horrific terrorist attacks in New York and Washington only two days after the conference closed. Indeed, hate is the same ideology that connects Durban and the terrorism of September 11, and it is the same ideology that produced terrorists such as Osama bin Laden. It should be of great concern to the U.S. government that Egypt, Saudi Arabia, Pakistan and other allies, after years of American economic security and support, did not just support the effort to vilify Israel and subvert the agenda of the World Conference – they led it.'*

I had been so busy in the weeks after 9/11 responding to the rapidly changing human rights environment that some time passed before I recognised that I should have moved immediately to defend the record on Durban. Indeed, I didn't fully appreciate until I moved to New York, a year later, that I had become a kind of hate figure within some elements of the Jewish community. That I could be viewed in this way when my life has been dedicated to tackling issues such as inequality, discrimination, and racism was (and is), although I have learned to accept it, indescribably painful. I found it difficult to defend myself, I believe, precisely because the accusations were so groundless.

I heard, indirectly, that Kofi planned to put forward his close friend and colleague Sergio Vieira de Mello as my successor. Sergio was one of the UN's most experienced officials and, unlike me, was a UN insider, coming up through the ranks. I admired the role he had played in Kosovo and East Timor, and had visited him in both places. In Kosovo, Sergio invited me to join him at a ceremony in Pristina where he was appointing judges. He joked that it made him feel like God, and it was evident that he had close to god-like powers on behalf of the United Nations, which he exercised with a deft mixture of prudence and flair.

While I was disappointed not to be continuing in office, Nick was relieved. Indeed, he was positively beaming, understanding sooner than I did how our quality of life would improve. It still rankled with him that he was never allowed to travel with me on what were, in some cases, dangerous journeys, and he was looking

---

* Tom Lantos, 'The Durban Debacle: An Insider's View of the UN World Conference Against Racism' *Fletcher Forum of World Affairs* 26(1) (Winter/Spring).

forward to being able to stop worrying about my safety and well-being. Nick reassured me that I had stayed true to my principles in the face of powerful forces, and that this release from the burden of office was a blessing. My tendency, if I suffer a setback, is to look for the positive, and infected by Nick's relief and enthusiasm, I began to think positively about the next phase in my life and career.

Imagine my surprise, on filling a standard UN form in 2002 relating to my retirement, when Nick pointed to the small print, which read, 'Does your spouse work? If so, does *she* work in the United Nations?' Initially we both laughed, but then I decided to ask my gender team to compile a formal complaint from me to the secretary-general – he was, after all, the head of the secretariat. I had to tone down the vehement text the team submitted to me before sending it off, and I never did receive a response from Kofi's office.

Sergio Vieira de Mello did, indeed, succeed me as UN High Commissioner for Human Rights in September 2002, only to be tragically killed less than a year later, along with twenty other UN staff members, during the Canal Hotel bombing – a suicide bomb attack on the UN headquarters in Baghdad in August 2003. Sergio was a man of immense charm and talent, and his loss was keenly felt throughout the organisation.

At the time of the attack, Sergio had been with another friend of mine, Arthur Helton, who was killed alongside him. Arthur was an expert on migration and a popular lecturer at Columbia University in New York, with a dry sense of humour. He had done a great job moderating a conference session of the American Bar Association in San Francisco I had spoken at, at his invitation, only ten days before the bombing in Baghdad that claimed his life. We sat beside each other at the formal dinner that evening, and Arthur told me that he would be going to Iraq to discuss migration issues with Sergio, and promised to pass on my warm regards and to bring back the real gossip.

Some weeks later, in a hospital in Oxford, I visited Arthur's good friend Gil Loescher, who had survived the bombing but lost both legs. The nurses were full of praise for his resilience and spirit. We spoke affectionately about Arthur, and how he had promised me the gossip from Sergio. I was very touched when Gil used his badly injured hand to inscribe a copy of his book on migration for me. His daughter Margaret made a moving documentary, *Pulled from the Rubble*, about how Gil and their family coped. We mourn the dead, but too often we forget the devastation caused to badly injured survivors and their families.

I was certain I wanted to carry my work forward, but in a more flexible way, free from the gruelling bureaucracy and the often dispiriting sense that despite the rhetoric, the international community wasn't taking human rights seriously enough in conflict situations or in tackling poverty and promoting economic development.

My instinct was to work out of the United States, and to find a forum to remind that country of its own high ideals and the importance of its earlier pathfinding on human rights under inspirational leaders such as Eleanor Roosevelt and Martin Luther King Jr.. I wanted to continue to challenge the post-9/11 climate and the knock-on consequences in other countries. I felt connected with the States, both because of the hugely formative impact of my time at Harvard and because of the strong friendships I had built up over the years.

I decided to base myself in New York because it was the head-quarters of the United Nations, with which I wanted to stay in touch, and an international city from which I could carry out my new agenda: encouraging the United States in its international human rights commitments, and pioneering practical work on economic, social, and cultural rights, and the empowerment of women, with a focus on the impact of globalisation on African countries.

As the end of my term approached, I received some enticing job offers. Stanford University proposed to create a chair, with the idea of having a centre within the university. Nick was keen on this offer from a world-class university, and we had close friends living in Palo Alto. But I persuaded him we should turn it down. It was too far away from the United Nations and from our family in Ireland, and even farther from Africa, the continent I was most interested in supporting as a human rights advocate.

As September 2002 drew near, I made connections with potential allies. I accepted an invitation to become honorary president of Oxfam International, an organisation I had long admired, especially since, a few years earlier, Oxfam had committed to a rights-based approach in its work in development.

In mid-2002, I entered into an agreement with the then CEO of the Aspen Institute, Elmer Johnson, who was enthusiastic about a new programme that would be based at the institute and would address international human rights. The Aspen Institute, located in Washington DC, had a reputation in the United States as an organisation still seeking to reach across political and ideological divides for meaningful dialogue. I would have space to shape the programme in any way I chose. I didn't want my work to be simply another programme of the institute; rather, I wanted Aspen to be one of the key partners in a new civil society initiative I would set up, with a group of colleagues, linking ideas of human rights, and development.

I was fortunate that John Healy, CEO of Atlantic Philanthropies, offered generous financial support for what he termed 'thinking time', to allow me to bring together a group of advisers and develop a new initiative, concentrating on how to be entrepreneurial with human rights, not just emphasising adherence to norms and standards but actually changing people's circumstances on the ground. Like many others, I learned a great deal from the rigorous approach of Healy and the quiet philanthropist behind Atlantic, Chuck Feeney, whose refreshing attitude to the disposal of the fortune he had earned was epitomised by his assertion that however wealthy you might be, you could only wear one pair of shoes at a time.

I was buoyed up by a strong sense from my colleagues in the United Nations that my role as High Commissioner had been appreciated and that there was genuine regret that I would not be continuing in office. From the beginning of my tenure, I had set myself a personal goal: to leave the Office in much better shape than I found it. When I looked around at my colleagues, I saw a bright, cohesive, and committed team working together on this often thankless task with strong morale out of our new home in the Palais Wilson in Geneva, our collective expertise advanced.

On my last visit to New York, Kofi had held a reception in my honour, gathering together a number of human rights friends as well as UN officials. By that time I had got over any personal hurt and had mentally moved on, having had several good meetings with Sergio Vieira de Mello to brief him and ensure a smooth transition. I was touched, not just by the good will, but by the recognition of all the hard work that my team and I had put in. In Kofi's own words:

> The job of United Nations High Commissioner for Human Rights is not for the faint of heart. At times even well-behaved Governments view the occupant of the post as something of a nuisance, while those with something to hide will often denounce the High Commissioner's efforts as unwarranted attacks on national sovereignty. Civil society organizations, meanwhile, often expect miracles – as though the policies of hard-bitten dictators could be changed overnight by confrontational public comments, or for that matter by hidden persuasion, from an official whose power is entirely of the 'soft' variety. And as if this wasn't enough, the High Commissioner must also run a sizable administration and navigate the political minefields associated with the Commission on Human Rights and its wide-ranging mechanisms. Equal parts lawyer and teacher, prosecutor and witness, hard talk and soft shoulder, the job, though little more than a decade old, is one of the most important in the entire United Nations system.

Back in Geneva they threw me a lovely send-off party, presenting me with a book of photographs of various events, which they had all signed with messages for the future. It was difficult to say goodbye to these people with whom I had worked so intensely over the previous five years. There were many hugs and some tears.

My last working day as High Commissioner was 11 September 2002, the first anniversary of 9/11, and my final event was to speak at an interfaith memorial commemoration organised by the dean of the Cathedral of Geneva, William McComish. I recalled where Nick and I had been on 9/11, our trip to the Inishkea Islands off County Mayo, and then chose to read Seamus Heaney's poem, 'From the Republic of Conscience' (the final stanzas of which are set at the beginning of this chapter), as the last official words I would speak as High Commissioner.

With these words I reminded myself, Mary Robinson – private citizen of tomorrow – that 'no ambassador would ever be relieved'.

17

Realizing Rights in Practice

*The power of human rights is the idea that all people are born equal in
dignity and rights. But it's an idea that doesn't have an enormous army
behind it. It's an idea whose effectiveness depends on people themselves believing in it, taking it up, and being willing to defend it.*

Dorothy Q. Thomas, in Worden, ed., *The Unfinished Revolution*

THE NEXT morning, Nick and I left Geneva and drove to the farm-
house of friends in the Luberon in France to meet with my brother
Adrian, his wife, Ruth, and some of our young. Nick and I badly
needed a holiday. Even as we drove I could feel the tension lifting.
Our car was packed to the hilt with our remaining belongings, the
rest having been sent ahead to New York. As we crossed the border
from Switzerland and into France we both let out a whoop of relief,
and then started to laugh. A new phase in our lives had begun.

We went for long walks in the sunshine, flanked by fields of laven-
der. We swam in the pool, and read, and dozed in the warm shade.
We ate outdoors and drank delicious local wine; we relaxed and let
our hair down. Of course, Nick and I discussed what the future
might bring, both with each other and with our friends and family,
but these were relaxed and light-hearted discussions, with plenty of
teasing and satire and self-mocking. We were happy to laugh and
have fun, not to tax ourselves too much yet. For the first time in
years we had time on our side and were not slaves to deadlines or
having to react instantly to world events. My relief at no longer
carrying the burdens of High Commissioner was evident to all of us,
and Nick noted that I looked years younger and that my sense of
mischief had returned. At one of our family dinners, Nick and I
donned enormous fancy hats that had belonged to the grandmother
of the house, causing much amusement.

Two weeks later Nick and I flew to New York to start our new lives. We fed off the energy and eccentricities of this city always on the move. There is something special about New York, a spirit of urban energy, where the population is constantly reinventing itself in pursuit of its dreams. We adapted to our new lifestyle and quickly established some essentials in our neighbourhood: where to find a good meal, where to bring our laundry, how to use the subway. When I bumped into other Irish people in the city, whether tourists or people working there, they would look with surprise as they recognised me in a crowded subway car or walking around Union Square and would then approach me with some concern to ask if I was okay and if I knew where I was going!

The Aspen Institute had organised a modest office on Madison Avenue to house our new group, so while Nick dealt with the day-to-day matters, unpacking our belongings and setting up home in a new city, I began to organise our New York office, which would be linked to Aspen's headquarters in Washington DC, where other colleagues would be based. I had persuaded my invaluable executive assistant, Cecilia Canessa, to take a year's leave of absence from the United Nations to help me make the transition. Scott Jerbi, who had worked closely with me in the Office of the High Commissioner, had agreed to join me in the new initiative and work out of the offices of the International Council on Human Rights Policy in Geneva.

I fell into a busy schedule of speaking engagements for the final months of 2002, including at Harvard, Yale, and Columbia universities, and that dissolved any concerns I had about finding a platform to express my views. These, and subsequent invitations, enabled me to speak out on the two distinct agendas I was pursuing: the way counterterrorism measures were tending to undermine human rights, and tackling poverty as a human rights issue.

When the dean of the School of International and Public Affairs at Columbia University, Professor Lisa Anderson, invited me to teach a graduate seminar on human rights and globalisation, I was delighted. I love teaching. For me, teaching has always been a way of learning, and I learned a great deal from the bright, resourceful students from around the world who had come to Columbia and selected my course. Most of them had experience of working in developing countries, and they helped shape my ideas as we discussed how to implement economic and social rights in practice on the ground.

Having been given 'thinking time' by Atlantic Philanthropies, I took just that. I felt liberated to be able to act independently again and to work out, with my colleagues, our own agenda. This was an opportunity to start again from scratch, to develop ideas on how economic and social human rights could be promoted.

The expression '*human rights*' carries different meanings and resonates differently in different parts of the world and within countries, depending on political systems, cultural influences, religious views, and, importantly, economic development. The Universal Declaration proclaims the fundamental rights, which were also encapsulated in the 'Four Freedoms' State of the Union address by President Franklin D. Roosevelt in 1941:

> The first is freedom of speech and expression – everywhere in the world. The second is freedom of every person to worship God in his own way – everywhere in the world. The third is freedom from want, which, translated into world terms, means economic understandings which will secure to every nation a healthy peacetime life for its inhabitants – everywhere in the world. The fourth is freedom from fear, which, translated into world terms, means a world-wide reduction of armaments to such a point and in such a thorough fashion that no nation will be in a position to commit an act of physical aggression against any neighbour – anywhere in the world.

I had been frustrated as High Commissioner with the failure of industrialised countries to take seriously 'freedom from want'. I felt that more priority needed to be given to connecting the human rights agenda to the reality of people's lives as they coped with situations of conflict and poverty.

In my speech in Oxford at the beginning of my term as High Commissioner in 1997, I had declared that 'poverty itself is a violation of numerous basic human rights'. I had in mind the basic rights to food, safe water, health, education, decent work. I was quickly corrected in my use of the term '*violation*'. 'Violations' are committed by states. Strictly speaking, poverty cannot 'violate' human rights. But my point was that poverty undermines human rights to a devastating extent.

The focus of our work would be on how to make a real difference in advancing the human rights agenda as part of broader debates on trade and development and poverty alleviation, in particular in Africa, the part of the world that had been most marginalised by increasing globalisation throughout the 1990s.

From the outset, Realizing Rights (as our initiative quickly became known) was not intended to be permanent. Our aim was to develop projects that would have long-term impacts outlasting our involvement. We stressed the double meaning of 'realize': Everyone should realise that he or she has a birthright of human rights – all countries have signed up to the Universal Declaration on Human Rights – and those with power (governments, international bodies, and corporations) should realise those rights by implementing and respecting them.

Over the early months of 2003, during which we planned our initial strategy and priorities, we kept coming back to the question of where my own strengths and influence could best be put to use. As a former head of state and former UN High Commissioner, I had access to heads of state and government, and leaders of civil society, business, and academia. I had always stressed the importance of grass-roots organisations and local communities having a voice on decisions affecting them, and had retained some useful contacts from within such groups. The dialogue on issues would have to be both top down and bottom up. Through such 'asking power', Realizing Rights could bring senior government and UN officials to the table with those who were marginalised or working at the local level and in need of access to policy makers to have their voices heard.

One of our areas of involvement, under the leadership of Heather Grady, was in 'decent work'. Heather had worked at a senior level for Oxfam in Asia before joining Realizing Rights, bringing that organisation's commitment to a rights-based approach to its work. Taking on the role of managing director, she brought to our small catalytic organisation not only expertise on development issues but also integrity and a work ethic that set a high standard for our programmes as a whole.

Our starting point was the 'right to work' under Article 23 of the Universal Declaration of Human Rights: 'Everyone has the right to work, to free choice of employment, to just and favourable conditions of work and to protection against unemployment.' But 'decent work' was not commonly considered a human rights issue by activists, and was thought of more as a labour issue. We linked to the International Labour Organization's approach to 'decent work': work that is productive; offers a fair income and social protection; provides opportunities for personal development, and promotion, and organising and participating in work-related decisions; and ensures equality in the work-place.

Where you have huge numbers of poor people, particularly young people, who are unemployed, their sense of identity and self-worth are seriously affected, and very often there is huge pent-up frustration, as evidenced recently in the Arab Spring in Tunisia and Egypt, which spread to other countries in the region. The high number of educated young unemployed bursting with frustration in those countries wanted not just jobs but also recognition of their dignity and worth. They demanded participation and, through social media, which allows for collective participation and organisation, tackled the dictatorships and corruption that were repressing their essential freedoms.

Our embrace of 'decent work' sought to make the case to human rights advocates that this issue wasn't just for trade unions or development experts but required the energies of wider constituencies calling for democracy, good governance, and respect for human rights. We wanted to show that putting more emphasis on encouraging work opportunities in poorer countries – and not just any work, but sustainable, productive work that came with social protection – could impact positively across societies.

But how to turn these grand ideas into practical policies and programmes that could help people in their daily lives? Much of our early efforts were experimental and designed to demonstrate the value of new and more holistic approaches that drew on international human rights standards and commitments. We began working at the country level.

We focused, for example, on Liberia, which was emerging from civil conflict, working with the minister for labour, Kofi Woods, who was a human rights activist before his appointment to cabinet. He rejected a strategy that centred on attracting investment through the availability of the cheapest labour, saying from the outset: we have standards, we want decent work and equality of pay for men and women. He got little support from the traditional donors, whose view seemed to be that a very poor country such as Liberia should indeed offer the cheapest possible labour, should, essentially, take what it can get.

One project was a road repair scheme supported by the International Labour Organization, in which both men and women were trained in maintenance of rural roads, including constructing and inserting drainage pipes so that, during the rainy season, flood water could run off. It was instructive to learn how receiving equal pay and having the opportunity to move to a supervisory grade on the same terms as men enhanced the morale and status of the women involved.

When I talked to Kofi Woods it was clear that he was frustrated at not being able to assess the size of his workforce, or the number of unemployed in his post-conflict country, where thousands had had to flee fighting. 'How can I make plans, Mary, when I can't even measure the size of the problem?' he asked.

Despite the fact that we were not a grant-making body, Realizing Rights made an exception and provided Kofi Woods with modest funding to carry out a labour force survey. He was then able to leverage this amount and get larger funds, and my colleagues helped facilitate the process and expertise that led to the drafting of new legislation providing for equal pay and other safeguards.

The move to New York meant that life was much better for Nick. He was again able to take an active interest in the work I was doing, and came with me to various events. He was also keen to be in

Ireland to do further research and writing on the history of caricature and on conservation projects, and to attend to our home in the
West of Ireland. Though he insisted that I not work too hard so that
we could spend some weekends together enjoying New York, my
work continued to mean that a fair proportion of my time was spent
travelling either within the United States or abroad.

Somehow these are not real separations, because on any day that
we are not physically together – and this has been true of our entire
married life – I phone him to touch base. The reason it's usually me
calling him is that I am often harder to reach. Sometimes I get the
time zone wrong and call him at three in the morning his time. He
generally laughs and says, 'It's okay. Always ring me. I don't mind
being woken up.' It would be difficult for me to be out of touch
with Nick for more than twenty-four hours. These phone calls are
an important way that we share the partnership of life together.
Quite often I talk over something I am about to do or tell him about
something that I have done. And, of course, he remains that special
friend who tells you the things you need to hear, not the things you
want to hear. There have been many, many times when he has said
to me, 'I don't think you are going about that the right way,' or 'Why
are you doing that?' Inevitably I would see that he was more than
half right and adjust my approach accordingly.

I have seen and experienced so much of what goes on in the
world through my work, but it is always with great relief that I land
back with Nick at whatever place is our temporary home. Better
still are those occasions when we can go to our beloved home in the
West of Ireland to be with our family. There I can really recuperate,
going for long walks in our grounds, breathing in the clean air, and
visiting familiar trees as old friends, and some newly planted.

I was thrilled when on one of these walks Tessa told me that she
and her partner, Robert Purcell, were expecting a child. I was to
become a grandmother! To have your daughter giving birth is
extremely special. It gave me an extra spring in my step as I went
about my work. Seeing her bump showing a few months later
brought back memories of my own pregnancies. On the morning
of 12 December 2003, which was our thirty-third wedding

anniversary, Nick and I had the pleasure of visiting Tessa with little baby Rory in her arms and Robert looking on as the very proud father.

I still remember the remarkable adrenaline rush that I had at the idea of a grandson, this Rory, and a complete sense of shift in my own time horizon. From now on I would be thinking at least a hundred years hence. It reinforced my increasing preoccupation with what the world would be like in 2050. Rory would be in his forties then. How was it possible to make sure that we made the right decisions so that he, and all the children born in this year, 2003, would not find that their world was a conflicted and terrible place because of the mistakes we were making?

In March 2005, Rory's little sister, Amy, was born, and I was now the proud granny of two. I would take every opportunity to produce photographs to show my friends and colleagues, wherever I was in the world. My brother Adrian, still living near Ballina, his solicitor's practice now housed in our childhood home, Victoria House, was enjoying his own set of young grandchildren. The two of us would revel in this new generation when they got together in one of our homes in Mayo, once more, happily, full of babies' toys.

William, practising architecture, had moved to Barcelona, and in early 2005, while Tessa was still pregnant with Amy, and Rory was taking his first steps, the family travelled there for William's wedding to his partner, Bibiana Macias. It didn't take long, then, for our Spanish family to grow. When their first born was due, I was determined to be there to welcome this third grandchild. With great advance planning, I presented myself in Barcelona three days after the due date, to find a very pregnant Bibiana and a rather apprehensive William. I stayed for a few days, taking them out and trying to help them pass the time, but it was not until after my departure that baby Otto arrived, in June 2006. His little sister, Kira, followed in March 2009.

Aubrey, meanwhile, had moved to Cape Town in South Africa, where he was continuing his studies and becoming involved in a local NGO, Mamelani, focused on the plight of street children. I managed to get to South Africa quite regularly, which allowed me to catch up with Aubrey. I would take him out to one of his favourite restaurants and the two of us would have intense discussions about what was happening in the world.

Realizing Rights developed a health programme, working with the governments of Ethiopia, Sierra Leone, Mali, and Senegal to address their responsibility to 'progressively realis[e] the right to the highest attainable standard of health'.* This programme was led by Peggy Clarke, who had been vice president for programmes in the Aspen Institute and helped me found Realizing Rights.

The classical human rights stance is to identify violations and 'name and shame' when governments fail in their responsibilities. We advocated a new approach to implementing economic, social, and cultural rights that took a more development-oriented line by saying, yes, you have to criticise governments where appropriate, but you also have to be prepared to work with them in a collaborative way, – as the jargon puts it, – to build their capacity.

Essentially we helped health ministers and their senior officials from Ethiopia, Sierra Leone, Mali, and Senegal implement their own policies for reproductive health and family planning – an issue I remembered well from my own days as a young legislator – and develop health insurance schemes so that health problems would not drive poorer families into further poverty. The officials from these four countries welcomed opportunities to come together with professional health experts, and they found this peer learning, sharing one another's experiences, both effective and enjoyable. Our special adviser for this work, Dr Francis Omaswa from Uganda, who had been secretary-general of the Ugandan health ministry before moving to Geneva for a period as executive director of the Global Health Workforce Alliance, spoke passionately about the way the International Monetary Fund and the World Bank's structural adjustment policies had weakened African health systems (cutting their staff), which led to a rise in maternal and child deaths. 'African governments them-selves,' he said, 'need to give health greater priority in their budgets.'

---

* International Covenant on Economic, Social and Cultural Rights.

Another issue Francis was passionate about was the dramatic increase in health worker migration from some of the poorest countries in Africa, which resulted in weakening those countries' fragile health systems even further. Health worker migration is the common practice whereby doctors and nurses are wooed away from their own countries by the higher salaries and better conditions offered in richer countries. Richer countries would point out that they are doing these individual doctors and nurses a great favour, offering them better work opportunities, but this ignores the fact that, in the process, they are depleting the health systems of poorer countries – (who had educated the health workers) – of this valuable resource.

Over a three-year period Realizing Rights worked with the World Health Organization to bring together an international advisory council that included ministers of health from a number of the affected developing countries, ministers of development from countries that recruited health workers, representatives of the medical and nursing professions, and public health experts. The objective was to draft a code of practice on the international recruitment of health personnel that would discourage active recruitment of health workers from developing countries.* On the day the code of practice was adopted there was clapping and cheering in the World Health Assembly in Geneva. Francis, Peggy, and I hugged one another with glee.

While my work rate after moving to New York remained tough, the whole office ambience was such a pleasing contrast to the bureaucracy-heavy atmosphere of the United Nations that it felt like a different world. My colleagues and I would often share 'brown bag' working lunches and, depending on what the afternoon held, I would quite often substitute a bottle of white wine for the standard iced tea, to inject a little more relaxation and fun. As we clinked glasses, we relished how sinful this would be considered in many of the neighbouring New York offices. And, of course, sharing a glass of wine made us more inclined to loosen up and tell personal stories, strengthening our bond. Heather liked to joke that this was our own version of 'decent work'.

★　　★　　★

---

* Eventually adopted in May 2010 by the World Health Assembly.

Classic human rights work involves ensuring that *governments* meet their obligations. Human rights treaties are drawn up by governments for governments. In my legal career, when I was taking cases to the Strasbourg and Luxembourg courts, I was bringing claims against Ireland for breach of its obligations under the European Convention on Human Rights or the EU Treaty, directives or regulations. My work as High Commissioner for Human Rights predominantly involved holding governments accountable for violations of human rights treaties. But there is another global aspect to international human rights: a growing recognition that corporations and the business community have a role to play and responsibilities, beyond workers' rights and labour law, in safeguarding human rights.

For six years, from May 2003, I chaired the Business Leaders Initiative on Human Rights (BLIHR) made up of a small group of leaders of multinational companies who were committed to assessing their responsibility to respect human rights standards, particularly in the developing countries in which they were working. Modelled on an existing business leaders' initiative on climate change, the BLIHR comprised companies who were pioneers in their willingness to discuss their human rights responsibilities.

Chairing these meetings, supported by my colleague Scott Jerbi, I learned what motivated business leaders to engage with human rights. Their first thought was to assess and minimise their risk: they did not want the business's reputation, its brand, to be damaged, as had happened to the global sportswear company Nike when it was revealed, in the 1990s, that child labourers in Pakistan were working in sweatshop conditions to manufacture its soccer balls. To protect their reputation and brand, multinational companies such as Nike increasingly realised the need to insist on higher standards for workers in the supply chain, to bring them in line with international labour and human rights obligations. This might entail, for example, raising the minimum age of employees, or imposing greater standards of health and safety protection, or providing for monitoring of conduct throughout the supply chain.

BLIHR companies soon discovered that this engagement with human rights was actually very important to their employees as well;

the boost in corporate morale could help them retain highly quali-fied staff. A further benefit was that businesses, especially large corporations working in poorer, developing countries, found that when they engaged in the human rights dialogue and familiarised themselves with their appropriate responsibilities, they were better able to respond to and contain the damage if a problem emerged in the supply chain or elsewhere.

An example is the case of The Gap clothing company. The Gap was not one of the founding members; it joined BLIHR after the clothing company issued a social report in 2004 acknowledging that it had had problems in its supply chain. These problems had led to public criticism and even some picketing of its stores for sourcing products from factories in developing countries with bad labour practices.

The Gap's representatives at BLIHR meetings were enthusiastic members, but they warned that it is extremely difficult to guarantee there will be no problems when products are sourced from poorer developing countries. As one Gap representative put it, 'The cheaper the product for the consumer, the more likely it is that there were abusive work conditions where it was manufactured.' So it proved, quite dramatically, for The Gap. In 2007, the *Observer* newspaper in the United Kingdom broke a story about children in India working in conditions close to slavery to make Gap clothes. CNN took up the story, asking its viewers if they would continue to buy clothing from the Gap, given the terrible sweatshop conditions under which the garments were made. A large majority of viewers indicated that they would not.

The Gap had to move quickly to try and diffuse the situation, and this it was able to do as, since its audit in 2004, it had a code of conduct already in place to stop the abusive conditions, and a number of employees around the world dedicated to ensuring compliance with it. While the policy had not succeeded in this instance, senior representa-tives made it clear that The Gap had been unaware that products had come to it through a long pipeline from that factory in India; that it had destroyed the goods in question, and that it had sent representatives to the factory to see if training and safeguards could be put in place to remove the abusive conditions while preserving local jobs. By being proactive in respecting human rights locally, including the right to work, the Gap was in a position to defend its brand.

Meanwhile, the United Nations had created a new position to clarify the responsibility of business for human rights, and Professor John Ruggie, who had previously helped Kofi Annan develop the UN Global Compact for business, trade unions, and civil society, was appointed. Somewhat controversially, and causing a certain tension within the human rights community, Ruggie took a new approach, and instead of starting from the perspective of victims of human rights abuses involving corporations, he started to build a framework that business leaders themselves would be involved in developing and support as reasonable and fair.

Initially I was critical of this approach, but I soon realised that its rigorous, analytical method was engaging business at a serious level. Realizing Rights, led by Scott Jerbi, became a bridge between John Ruggie and a sceptical human rights community, and an ally to John, working with him to clarify the responsibilities of all companies for respecting human rights principles. His work was hugely important in clarifying the distinction between the responsibility of governments to protect human rights and the responsibility of corporations to do the same, with the need to provide better remedies where breaches occur. The guiding principles of 'protect, respect and remedy' were endorsed by the UN Human Rights Council in June 2011. I wrote to John congratulating him and his team on 'a good day's work accomplished in six years!' His response was typical of his wit: 'As Bismarck said, if you like sausage, don't watch how it is made.'

I continued to pursue my other broad agenda by giving speeches and delivering lectures on the rule of law and the importance of governments upholding human rights even when they were increasing security measures to combat terrorism. I was a vocal opponent of the Iraq War, which I argued could not be justified under international law. I was heartened, later on, when participating at meetings of the American Bar Association on rule of law issues, to find that the majority of its members were of the same view.

When the International Commission of Jurists (ICJ), of which I was a member, asked me to serve on an 'eminent jurists' panel' on counterterrorism and human rights, I readily agreed. The panel, chaired by Arthur Chaskalson, former Chief Justice of South Africa, a distinguished jurist and wonderful human being, conducted public hearings on counterterrorism in different regions over a three-year period. The panel had been the brainchild of a friend and former UN colleague, Nick Howen. Shortly after the establishment of the panel, I succeeded Arthur as president of ICJ, and some months later Nick began a brave but losing battle against bone marrow cancer. His untimely death caused great sorrow in the human rights community.

Our panel report, 'Assessing Damage: Urging Action', released in 2009, was well received, even by some of the governments we had been critical of during the hearings process, such as the United Kingdom and the United Sates. We promoted it publicly and privately in different regions of the world, and it was discussed at a meeting of the UN Security Council. In many ways the report became a focal point in a reassessment of counterterrorism measures. Needless to say, it also provided me with fresh material for my own advocacy on the importance of upholding international human rights standards while taking effective measures to counter terrorist attacks.

From the beginning, I was determined that a significant strand of our Realizing Rights work should involve support for women addressing the terrible violations and sexual abuse suffered in situations of conflict and post-conflict. Unlike when I was first elected to the Irish Senate, I now have no difficulty describing myself as a feminist, although many feminists would describe me, instead, as a woman activist who is supportive of women's empowerment and women's movements organising to achieve social change but not a true feminist in their sense, whose activism is specifically directed towards challenging and changing women's gender subordination. Over the years, I have come to appreciate the value of the work done by the feminist movement, and how feminist discourse influences the character of speech and expression in the home and the workplace, and at times in politics, culture, and the arts. Many women activists, like me, call ourselves 'feminists' precisely because of the term's strong affirmative connotations.

Our plan was to engage women in positions of influence who had access to power to travel to parts of the world, especially African countries, where women were suffering from such violations, with the aim of getting their voices heard so that effective measures would be taken to protect them.

My experience, both prior to and in the United Nations, led me to conclude that it is no longer the case that women are excluded from positions of power. Now it is more a question of how they are willing to exercise power. Women's groups have pointed to a recurring problem whereby some women, even those from an NGO or activist background, when elected and given a ministerial role, change their stance and become part of a male-dominated team. When women achieve political power, they do so in male-dominated situations, and often the price they pay for that power is they don't rock the boat. Though it is terribly disappointing to see this happen, attitudes are changing, and this is becoming less and less the case. A new generation of women leaders is beginning to emerge, both inside and outside elected politics, with a clear sense of purpose, challenging male dominion with energy and vitality.

While there is a big concern, as I write, about lack of leadership on key global issues – the international financial crisis, the risks of nuclear proliferation, climate change, the failure to deal effectively with Iran or North Korea or with crises in countries such as Syria – I have a sense that an exception to this is in the area of women's leadership. I am not alone in this view. Graça Machel, the former minister of education in Mozambique, said in an interview in October 2011, 'I think that in ten years time Africa will be a completely different landscape. It's already happening with regard to women: skilful and ambitious women will be at the highest levels of decision-making in politics, business, science and technology. There's a new generation of female leaders coming.'

Women's leadership is coming into its own and is reaching a tipping point. It is happening at a global level, at a national level, and at a grass-roots level. Society as a whole benefits when women reach their full potential. As Mary Wollstonecraft, the pioneering author of *A Vindication of the Rights of Woman*, put it over two hundred years ago, 'I do not want women to have power over men but over themselves.' The challenge remains how to encourage women, when elected or appointed to political office, to retain a strong awareness of the need to surmount barriers, fight discrimination, and, from positions of leadership, empower women.

The question I faced was how, with the support of my colleagues Jennifer McCarthy and Michelle Bologna in Realizing Rights, could I use my influence and any convening power I might have to pioneer innovative ways in which women could exercise power in that sense.

Many women's organisations operating at a local level were frustrated at not being able to get resources for their priorities, which forced them to contort what they were doing to fit donors' fashions in funding. I heard women complain, for example, about being required to engage in work on HIV/AIDS. While they acknowledged that this is important work, what they really wanted to do was to build up their own empowerment and capacity to change the power dynamics in their communities. But very little funding was available to promote the issue of women's leadership at the local level as such; rather, women were encouraged to engage in issues that the donor community wanted to support.

I began to think more and more about how to connect the increasing number of women with power and influence – whether ministers; heads of state or government; business leaders; or women in leadership positions in UN agencies or elsewhere – with the millions of women who feel they are second-class citizens in society, who have so many barriers, so many inhibitions, are so often undermined, and so often live in poverty. It seemed important to develop a new type of women's connection that expressly linked women who had access to power and influence with those who, working at the grass-roots level, were also showing great qualities of initiative and responsibility.

I got to know excellent women's organisations based in the United States that were doing work internationally, such as the International

Center for Research on Women, whose leader at the time was Geeta Rao Gupta, and the Women's Learning Partnership, founded by Mahnaz Afkhami. I became close friends with both women. While the three of us are from different backgrounds and cultural experiences, spanning India, Iran, and Ireland, we quickly found common ground and developed a deep respect and affection for one another. We began to build a broad women leaders forum; and soon, for practical reasons, Realizing Rights took over and developed the forum as an internal programme, styled the Women Leaders Intercultural Forum. We believed in a different kind of transformative leadership, one that reached out and listened to women at the grass-roots level, and then brought their ideas to those who could exercise power to help them address their problems.

In November 2008 we tested this approach, bringing together a group of women leaders to carry out a supportive mission to Eastern Chad to help women and their families fleeing from Darfur and those internally displaced by local conflict in Chad. Darfur was at the height of its terrible conflict, and thousands were streaming over the border to refugee camps in Eastern Chad. In situations such as this, most of those fleeing are women and children. I was heartened by the response of the influential women I asked to join the mission. Dr Ngozi Okonjo-Iweala, who had been finance minister in the Nigerian government and would become managing director of the World Bank was happy to join us, as were Dr Musimbi Kanyoro from Kenya, then secretary-general of the Young Women's Christian Association (YWCA) worldwide; peace activist Asha Hagi Elmi of Somalia; Bineta Diop of *Femmes Africa Solidarité*; and Angélique Kidjo, the famous singer-songwriter and activist from Benin. Three non-Africans – the former German minister of justice Herta Däubler-Gmelin, former White House staff member Jane Wales, and I completed – the group.

We flew to Paris and from there to the capital of Chad, N'Djamena. On the airport terminal bus in Paris, I overheard with amusement and admiration Ngozi on her cellphone instructing her husband on the tableware to be used for the family Thanksgiving dinner she would be missing.

In the sweltering heat we were flown from N'Djamena in a small plane to Oxfam's compound in Eastern Chad. From there we visited the official UN refugee camp. We sat and listened to the women there tell their stories, overcome by the sheer brutality of the tales that unfolded. The impact of their trauma was etched on the women's faces. I recall one woman in her late twenties, her head covered in a green scarf, talking about an attack on her village from the air by planes, and on the ground by men on horseback. Pulling her head-scarf over her mouth nervously as she discussed the more difficult parts of her story, she described how she grabbed hold of her twin babies and ran as fast as she could into the bush and kept running for quite some time. Eventually she decided she needed to know what had happened to her husband and older child, so she placed her twins on the ground breath a tall tree and tied a ribbon around the tree so she could find it later. She went back to the village where she found that her husband had been killed and her older daughter raped and killed. She herself was then caught by soldiers and raped repeatedly. Badly injured, she crawled back to where she had left her babies and carried them over the border into Chad and to the refugee camp. This was not an unusual story. Each of the women we spoke to gave us horrific accounts of violence and rape. Even in the camp they did not feel safe. Men in uniform – either official soldiers of Chad or part of the rebel forces causing disturbances in that part of Eastern Chad – would attack them when they went in and out of the camp seeking firewood.

As we listened to the heart-rending stories, Herta Däubler-Gmelin asked, with mounting anger, whether these cases were being documented as evidence for the International Criminal Court. The woman who headed the small NGO that was providing trauma counselling for these Darfuri women shook her head sadly. In the first few months, they had begun to document these accounts, but their funding had been cut, and this part of their work had to stop, as trauma counselling was more urgent.

The following day we went to a camp of internally displaced Chadian women and their families who had had to flee violence in a town about seventy kilometres away. This camp was less organised than the official refugee camp, with shacks built of tarpaulin, corrugated iron, and found materials and located in a swampy area. Some women were attending a literacy course alongside their children. They were determined to write their own names and claim the food themselves at the centre, not to have to rely on men, who often cheated them.

We flew from Chad to Europe, spending several days in different capitals advocating for greater protection for these women and greater resources for the organisations supporting them. While in Chad, I had met Carl Bildt, the foreign minister of Sweden, and he confirmed that European countries were considering putting an EU force in Eastern Chad to provide protection, but that there was some internal division on the issue. We saw a role for us in helping overcome these roadblocks. Our advocacy began in Paris, with the foreign minister, Bernard Kouchner. In Berlin we met with the minister of international cooperation, Heidemarie Wieczorek-Zeul. In London we had a lengthy meeting with the prime minister, Gordon Brown, at which he was joined by the minister for women and equality, Harriet Harman. We took to the airwaves and spoke about the women we had met and their desperate need for protection and support. Shortly afterwards the EU did send troops and police to Chad to help with protection around the camps.

This visit to Eastern Chad stayed in my mind as an example of a new and persuasive type of women's leadership. As women who had access to power and influence, we could get close to and amplify the voices of women suffering on the ground. We could bring their reality directly to political leaders, simply by having access.

Bineta Diop was so taken with our visit to Eastern Chad that she was determined that African women leaders like her had to take more initiatives. A year later she invited me to Addis Ababa with a number of others, including Nyaradzayi Gumbonzvanda, Musimbi Kanyoro's successor as secretary-general of the YWCA worldwide; and Ruth Messinger, president of the American Jewish World Service. Our mission was to support the Sudanese Women's Forum on Darfur that Bineta Diop was mentoring, a group of more than a hundred Sudanese women. On the first evening, Bineta asked me to meet with the standing committee, which had given leadership to the forum during its first year, comprised of six women – coming from Khartoum, from Juba in the south, and from the three provinces of Darfur – along with a representative from the Darfuri diaspora.

We met in our hotel and sat in a circle. I was about to engage in a rather formal discussion about the forum when Nyaradzayi gently intervened and said, 'Before we talk about this forum that's being created, Mary, why don't we just tell a little about ourselves so that we can get to know each other?' Catching what she was at – this was just the sort of thing my mother would have encouraged – I said, 'Alright, Nyaradzayi. Let's start with you.'

She described growing up in Zimbabwe in a conflict situation with poverty in the village, but said that her father had believed in her and had enabled her to be educated, allowing her to pursue a career and go on to take up a leadership role first with the United Nations and then with the YWCA. One after the other these women told their stories. The common thread was 'my father believed in me'. One of them, from a village in West Darfur, was the only member of her village to have gone on to third-level education. Another, the daughter of a former prime minister of Sudan, lived in Khartoum and was seeking to support women's empowerment. Another was a minister in the government of Southern Sudan. These women came from different backgrounds and different walks of life but their father's faith in them was what had enabled them to become leaders in their own context. I spoke last, and noted that each of them had benefited from their father's support, as I had myself. Meanwhile, I had learned a lesson in leadership from Nyaradzayi. Instead of immediately beginning a conversation on a substantive issue, it can deepen the relationship to take the time to get to know the people you are engaging with, and draw out common threads at a human level.

This Sudanese Women's Forum on Darfur was eventually registered by the Sudanese Government, and gave women a voice and visibility they had not had before.

Some months later, in my last year at Realizing Rights, it was Nyaradzayi Gumbonzvanda's turn to take the initiative, inviting me to bring a group of women leaders to her home country, Zimbabwe, to encourage women there to have their voices heard. Again the incredible willingness of senior African women to join this visit was heartening. Our mission to Zimbabwe involved having to measure carefully how to criticise the serious human rights violations of President Mugabe and his regime, while establishing a working relationship with Mugabe. This was necessary in order to achieve our objectives of supporting women's participation in the constitutional review process that was under way, and facilitating the conference of women that we would be joining.

This conference was organised by the minister for women's affairs, Olivia Muchena of the Zanu-PF party and by Sekai Holland from Morgan Tsvangirai's opposition party, who was a member of the Organ for National Healing, Reconciliation and Integration. Across the terrible divide of the conflict in Zimbabwe these two women had forged a friendship. They linked with the vibrant Zimbabwean Coalition of Women, and during our stay, women from the different political parties signed a pact to support one another in the empowerment of women in the future constitutional process.

Our group enjoyed a comfortable lodge outside Harare, having resisted – to maintain our independence – the invitation of the government to stay in the major hotel and conference centre in Harare. I was shocked at the deterioration in conditions in Harare – including frequent power blackouts – since I had last visited there as Irish President in 1994.

When we arrived at the centre for the opening of this women's conference, I was told that President Mugabe would receive me in the green room together with several of his ministers, including the two women ministers who had been responsible for the conference's organisation. When he greeted me, I reminded him that during my state visit in 1994 he had invited me to his old school and had spoken warmly about Father O'Hea, the Irish Jesuit who had taught him. President Mugabe's face lit up. 'I remember that well, and you'll be pleased to know that there is now an O'Hea Memorial Hospital here, just outside Harare.' This exchange, I hoped, would establish a reasonable base with President Mugabe, which was important, as my speech did contain criticism of Zimbabwe's human rights violations, and I didn't want an overreaction that could prejudice our mission.

To my surprise, President Mugabe stayed on the platform for the keynote speech I had been invited to give. When I reached the passage in my speech criticising human rights violations, I could hear an uncomfortable shuffling of papers behind me. Sometimes it's hard to balance the need to speak out on human rights violations with the need to maintain reasonable relations in order to achieve progress on particular goals, in this case the quest to prioritise the role and involvement of women in the new Zimbabwe constitution. I held my breath, hoping my comments had not so alienated Mugabe that he would shun our Zimbabwean mission. But happily President Mugabe *was* willing to see the women's group afterwards in the green room, though it involved his lecturing me for about fifteen minutes on colonialism and land rights issues in his country. However, he clearly saw the mission as being one that was helping Zimbabwe, and he met with our women's group again after we had gone on site visits around the country.

It was an extraordinary experience to work with this group of committed, strong, courageous African women. They came from different countries and each had had a distinguished career. Dr Brigalia Bam of South Africa was chair of the Independent Electoral Commission. Dr Achola Pala of Kenya was former head of UNIFEM in Africa. Dr Thelma Awori, of Uganda and Liberia, had also worked in the United Nations, and had served a few years earlier as UN Resident Representative in Zimbabwe. Ambassador Lois Bruthus was the Liberian ambassador to the African regional body Southern Africa Development Community (SADC) and former president of the Female Lawyers of Liberia. Elizabeth Lule from Uganda was a senior official at the World Bank and former manager of the World Bank's programme on HIV/AIDS in Africa. They referred to me affectionately as their African-Irish sister.

We worked very hard, getting up early in the morning and putting in a gruelling day either at the conference itself or making site visits. One evening we decided to dance together. The record player was found and, after our meal, we cleared a space, where we danced until the early hours. It was an opportunity to express our warmth and affection for one another, African and Irish style!

Our visit did not change the political landscape of Zimbabwe; the political divides were far too entrenched for that. However, one small but significant positive outcome was an agreement signed publicly on the final day by women representatives of different political parties. Another was the subsequent inclusion of the two women ministers and representatives of the Zimbabwean Women's Coalition in the twice-yearly meetings of 'Gender is my Agenda' (GIMAC), held on the eve of African Union summits, an umbrella group of some forty women's networks throughout Africa created to focus on women, peace and security.

Other women have been engaged in similar ideas of how to give leadership. It was no accident that once a few women had been awarded the Nobel Peace Prize, they banded together and formed the Nobel Women's Initiative in 2006. This group has been active in a similar way on peace, justice, and equality since.

Appreciation of the value to society as a whole of women's economic empowerment is gaining ground all the time. A UN High-Level Panel on Global Sustainability report in February 2012, 'Resilient People, Resilient Planet: A Future Worth Choosing', highlighted its importance.

> Any serious shift towards sustainable development requires gender equality. Half of humankind's collective intelligence and capacity is a resource we must nurture and develop for the sake of multiple generations to come. The next increment of global growth could well come from the economic empowerment of women.

Taking into account the full economic contribution of women means taking seriously the work of those in the informal sector – street trading, working in markets, rural subsistence farming – jobs done primarily by women, even though they don't own the land or gain the profits from it. One of my heroines is Ela Bhatt, founder of the Self-Employed Women's Association in India, who has persuaded the Indian Government to factor into its GDP the contribution of the informal sector. It is a huge component of the economy of India, but until relatively recently it was not counted. To do so increases awareness and a realisation of the need for better social protection for these workers, better market facilities, a safer work environment. This, too, is part of a 'decent work' agenda.

I look back on my time with colleagues in Realizing Rights with deep affection and appreciation. It was a period when, free from the responsibility of public office, I felt energised to try out new ways of making human rights matter in small places, and to speak out publicly about the way some governments were undermining human rights standards in the name of countering terrorism. I learned a great deal: about how a small organisation can act as a catalyst and convenor, can be nimble and flexible without heavy bureaucracy, but must partner well with others to be effective.

Though we had initially intended to limit it to a five-year commitment, before I returned home to live in Ireland, Realizing Rights finally wound up, after eight years, in December 2010, hosting its last event, under the title 'Making Human Rights Matter'. We took stock of our overall approach, which had been to plant seeds, to empower others, to help shape new approaches, and then to step back and let others take the lead and move forward. In true Realizing Rights style, we made sure that the event ended with singing and dancing.

18

Being an Elder

The Elders can speak freely and boldly, working both publicly and behind the scenes . . . they will support courage where there is fear, foster agreement where there is conflict and inspire hope where there is despair.

Nelson Mandela

AT SEVERAL stages my life has taken a new direction I have neither foreseen nor expected. It was to happen again in 2007, when I was invited by Nelson Mandela to join a group to be known as the Elders.

Over the years, I had given many talks and taken part in many discussions on leadership: women's leadership, political leadership, business leadership, grass-roots leadership. But the element of leadership that really fascinated me was moral leadership. People who personify that moral leadership for me, an authority whose power derives from the integrity of the values they have lived by, include Nelson Mandela, Archbishop Desmond Tutu, and the Dalai Lama.

In early 2007, I received an intriguing letter from entrepreneur Richard Branson enclosing material about a group he was helping to form, 'the Elders', and asking to meet me. I had met Richard briefly several times at conferences, but we had never before had an in-depth conversation. Sitting on a couch opposite me in my office in New York, he embarked on an enthusiastic account of how the idea of bringing together a group of global elders had come about.

The idea had been put to Nelson Mandela by Richard and musician Peter Gabriel, who were long-time friends, and who conceived of the idea after the failure of an earlier initiative, involving Mandela, to try to stop the war in Iraq by persuading Saddam Hussein to retire into exile. In that attempt, time ran out and the bombing – 'shock and awe' – started. But the experience persuaded Richard and Peter that a group based on the idea of traditional village elders should be put in place, independent and privately funded, that could be ready to act in situations of crisis. They went to Nelson Mandela with this plan. He was enthusiastic and agreed to invite people to join the group, which would include his wife, Graça Machel, an extraordinary woman in her own right, former minister of education in Mozambique and widow of Mozambican president Samora Machel, who was killed in an aeroplane crash in 1986.

Richard then explained that gatherings had been convened in different regions representing civil society organisations, academics, and philanthropists, who were asked to discuss the idea of global elders and come up with possible names that could be put forward to Mandela for his consideration. My name was one of them.

Somewhat sceptical of how this might work, but honoured to be included in the group, I agreed to attend a planning meeting in May 2007, to take place in Richard's game reserve in Ulusaba, South Africa. Richard assured me that spouses would be welcome and included in all meetings, a further attraction. The first group of Elders (Graça Machel, Archbishop Desmond Tutu, Jimmy Carter, Nobel Laureate Muhammad Yunus and I), sat at a round table, Kofi Annan joining us by live video link from Geneva. We all knew each other and settled into discussing our working methods and teasing out how we would determine our initial priorities. Archbishop Tutu clearly relished the job of chairing (and bossing) this gathering which included former presidents and a former UN secretary-general. He set an early tone of informality by inviting us all to call him Arch, and soon we were comfortable using first names around the table. In truth, we were all waiting for Mandela, who was due to join us at the second morning. His arrival was signalled by the local staff of Ulusaba singing their hearts out to welcome their beloved Madiba. He was Madiba to us as well, and that was how we greeted him as he was led in by Arch, on the arm of his wife, Graça.

Having spoken to us warmly and exchanged some teasing remarks with Arch, of whom he was clearly very fond, Mandela read a prepared note in his strong, distinctive voice, telling us we represented an independent voice, not bound by the interests of any nation, government, or institution; that we should act boldly, but listen to local people, as they knew the most about their problems; that we should reach out to the marginalised, and especially young people, but not claim to have all the answers. We should stress that every individual can make a difference and create positive change in his or her society. Listening, I felt any residual scepticism on my part vanish. I could see that there was scope for such a group of elders to have an impact, and each of us felt we were being given a shared responsibility to carry on Mandela's vision.

The plan was that, after the meeting, Mandela would go outside, followed by the Elders, for a photograph; then spouses, advisory council, and others would join us for more photographs. But it didn't work out like that, because as soon as Mandela went out and sat down on the seat provided, the local people swarmed around him, again singing. Arch laughed and threw himself down at Mandela's feet, followed by Jimmy Carter and me, then Richard and Peter Gabriel and others. We were all laughing, caught up in the joy of the occasion.

In the years since, I have learned a great deal from each of my Elder colleagues, which later included Martti Ahtisaari, former president of Finland; Ela Bhatt; Lakhdar Brahimi, former foreign minister of Algeria, whom I had known at the United Nations for his expert work on mediation and conflict resolution; Grö Brundtland, former prime minister of Norway and director general of the World Health Organization; and Fernando Henrique Cardoso, former president of Brazil.* We designated Aung San Suu Kyi of Burma an honorary member for as long as she was under house arrest, and kept an empty chair at our meeting table in her name, draped with silk.†

We were fortunate to recruit Mabel van Oranje from the Netherlands as our CEO; she put together a team of committed and talented people to support us. And under Arch's guidance, we forged

---

* Muhammad Yunus later retired because of his own work pressures.
† When Aung San Suu Kyi was elected a member of the lower house of the Burmese parliament in 2012, she stood down as an honorary elder.

a strong bond and group identity. Inevitably, with a group like this, differences of opinion would be expressed; there were tensions, for example, when we talked about how to deal with the Middle East, how to deal with South Sudan. Some of us were more cautious than others; some of us wanted more emphasis placed on involving civil society, women's groups, youth; and others urged quiet diplomacy behind the scenes. Tensions crept in also when it appeared that the analysis was coming from a narrow Western perspective. Then Lakhdar or Ela or Kofi would intervene and challenge that. From time to time, Arch chided us that we had to 'learn to elder', to try and emulate the wisdom and forbearance of the traditional village elders as they mediate family conflicts or disputes between neighbours. We began each session with a moment of silence, which Arch brought to an end with a brief prayer. This reminded us, somehow, that we were not a political think tank or organisation; that we were trying to bring to the issues we addressed a moral authority.

In my experience, leadership can be more effective when exercised with a lightness of touch. I enjoy using humour and teasing, often in a self-mocking way, to make myself more approachable. Over the years, I have learned a huge amount from Arch. I have watched the way he uses humour to break down barriers of awe about the fact that he is a Nobel Prize-winner or an archbishop or iconic figure, and how he makes people laugh, sometimes deliberately makes himself look and sound ridiculous. Then when people have laughed and enjoyed the joke, he comes in with his serious message. And he always has a serious message, even if it is only to affirm someone's goodness and the value of what they are doing.

Another important lesson, which I learned as a child watching my father and which the Elders emphasised, is the immense power of listening, a simple act and with enormous impact. I often think of Article 1 of the Universal Declaration of Human Rights: 'All human beings are born free and equal in *dignity* and rights'. That dignity of the self is affirmed by properly listening to people. What I really remember about events I attend are the conversations I have, looking into the eyes of somebody, drawing him or her out, finding common ground or a kindred spirit. I never remember what the furnishings in a room looked like – if there were curtains, or even fine pictures. I do not remember what people were wearing. What interests me is what moves other people, what pains them, what excites them. When I dance with my African friends, I find that we dance with our eyes as much as anything else; there is laughter and respect and warmth.

From our inception, the Elders sought to play some role in difficult conflict situations, but only if we truly believed our involvement would help the situation. Mandela urged us, at our original meeting in South Africa, to 'seek advice from expert organisations and work cooperatively with them . . . listen, bring together antagonists and protagonists, work with anyone who is motivated to resolve a problem, and give them support and the determination to do so.' When we decided that the Elders should become involved in a situation, we normally identified three or four of us to participate on behalf of the group, usually because of particular knowledge or interest these people had.

We don't pretend we can solve all the world's problems; there is a limit to what you can do in a given situation. What we have tried to do is bring our experience and collective moral authority to bear; to urge the values of reconciliation, of good governance, of respect for human rights, of equality and non-discrimination, of fair processes and even-handed administration of justice.

President Jimmy Carter has been involved over many years with North Korea, and assisted, in 2010, in the release of an American prisoner. In early 2011, with relations between North and South Korea at rock bottom, he received a message that North Korea would welcome a visit from him, together with other Elders, primarily to discuss improving relations between North and South, and with South Korea's key ally, the United States. We would soon discover that, for the North Koreans, the current food shortage was top of the agenda. The Elders' team engaged in extensive planning work prior to our visit, and quiet consultations took place with the key governments concerned: South Korea, China, the United States, and Japan. All official talks on North Korea's nuclear programme were at an impasse, so it was felt that a visit might help open up possibilities.

Jimmy Carter would lead the group, accompanied by Grö Brundtland (who had visited the country when she was director general of the World Health Organization), Martti Ahtisaari, and me.

We began our journey in April 2011 in China, meeting with the foreign minister and other officials. We explained our priorities and asked for their support, knowing that China was the nation with the most influence on North Korea. We also met with UN officials, who could speak to us much more openly there than in North Korea itself. One concern of Western donors was that food aid was being siphoned off to the army rather than being distributed to the wider population, but the UN officials were able to assure us that they were getting better access and therefore were better placed to supervise and monitor any food aid that got in to the country.

We then flew from Beijing to Pyongyang, where we spent several days meeting officials and visiting a food distribution centre, a hospital, and an orphanage. We drove along broad avenues, past monumental buildings, and saw teams of people sweeping the roads with brushes. There were flowers everywhere, both natural and artificial; the founder of the state, Kim Il-sung, had loved flowers, and growing them had become a national preoccupation, as had tending to the numerous monuments around the city.

One of the most depressing aspects of what we witnessed was the complete absence of independent thinking or questioning by the people, the lack of civil society in any sense. Our efforts to raise concerns about the impact of the food shortage on women and children, for example, were rebuffed by three senior women we met. Dressed elaborately in traditional costume, they refuted all our criticism, utterly denying there were any problems, but rather, treating us to a lecture on the paradise for women's rights they said existed in North Korea. There was a clear disconnect between what they said and the stark reality in front of their eyes.

I had to bite my tongue during this discussion. I wanted to interrupt, contradict, argue – but we had agreed not to be argumentative or judgmental. At the same time, it was apparent that some within North Korean officialdom, by bringing us to one of the food distribution centres, wanted us to see the reality of the situation. The entire population works for the state, and food is allocated to families from these distribution centres. At the time of our visit we were told that the calorie level per family was now alarmingly low – low enough to persuade this proud and generally secretive regime to seek outside help.

North Korea had recently suffered flooding and an outbreak of foot-and-mouth disease. This compounded the problems of a very primitive system of agriculture, with farm work done mostly by hand or occasionally by the oxen that were now suffering from the disease. A deep food crisis was impacting the population at large. A high number of children were suffering malnutrition and its side effects of stunted growth and development, which we witnessed on a visit to an orphanage. The children, who greeted us with smiles and curiosity at how different we were from them, quickly gathered around Jimmy, who has a natural ability, through his innate kindliness and magnetism, to cut through cultural or age barriers. Not all of the children, who looked slight for their age, with thinning hair, were orphans. Often, we were told, families left their children at orphanages when they could not cope with feeding them at home.

At the hospital, Grö, with a practical touch born out of her work at the WHO, took the lead in talking with workers. She was shocked at what she saw as a terrible deterioration in the conditions since she had last visited – there was, for example, no running water – and infuriated, given the budget spent by North Korea on military equipment.

Against this backdrop of poverty and severe food shortage, we were very uncomfortable to find ourselves being served five- or six-course meals, even for breakfast, at the luxurious guesthouse in which the North Korean authorities were hosting us. We discussed what we considered a misplaced priority of devoting such time and energy to cleaning the streets, tending to gardens, and celebrating the North Korean leadership while neglecting the welfare and basic needs of the people, but if as a group we wanted to focus on engaging with North Korea and encouraging an improvement in relations with the South, we would have to tread carefully and not jeopardise our mission by being too judgmental.

Our talks on improving relations, led by President Carter, were more hopeful. The North Koreans made clear to us that they were open to resuming talks on the nuclear issue, both with South Korea and multilaterally. This was highlighted in dramatic fashion as we were leaving. Our bus, taking us to the airport, was called back to the guesthouse so that a senior official could present us with a written message from Kim Jong-il. In it, he indicated that North Korea was ready to reopen dialogue on the issue – a message we conveyed to the South Korean authorities when we arrived in Seoul, and to the ambassadors of the United States, China, Japan, and the European Union.

Back in Seoul, we had an opportunity to meet some young people who had fled North Korea. With them, we could have franker conversations. Most of them had been separated for long periods from their parents, having left North Korea out of desperation, citing in most cases a lack of food. One twenty-year-old woman described how she had survived the famine of the 1990s as a small child. Then the failed currency reforms in 2009, which wiped out private savings, again led to terrible hunger. After thirteen years living with an aunt in North Korea, she finally fled, through China, to join her mother in Seoul. Another young woman told me that her mother had been captured in China and sent back to prison in North Korea. Another described how she had tried to kill herself by swallowing a needle.

At a press conference, we stressed the reality of the food crisis, and the moral imperative of providing food aid despite the lack of progress on the nuclear front. Shortly afterwards there were some positive outcomes: talks opened up again on the nuclear issue, a number of European countries provided more food aid, and the United States decided to resume supplies. The sudden death of Kim Jong-il, then, in December 2011, and the succession of his son, Kim Jong-un, inevitably changed again the dynamics of North Korea's relationships with these countries.

While our mission relating to improving relations on the Korean Peninsula was under way, Kofi Annan had been following closely events in Côte d'Ivoire and an ensuing crisis. Despite being defeated in the polls in November 2010, the outgoing president, Laurent Gbagbo, had refused to stand down, and fighting had broken out, including in the capital, Abidjan. Eventually, with the help of French troops, Gbagbo was captured, and the winner of the election, President Alassane Ouattara, took over in a very tense situation. Kofi felt that a visit from a group of Elders might help to defuse the situation. Initially the group was to consist of Kofi as leader, with Archbishop Tutu and Graça Machel. Unfortunately Graça was unable to participate. Because of the incidence of rape and sexual violence that had been reported in Côte d'Ivoire, and, indeed, to ensure a gender balance, it was felt that there should be a woman on the team. I had some experience of the region, having been in Liberia a short time before, where I met some women from Côte d'Ivoire intent on highlighting the role of women in mediating the conflict there. So I agreed at the last minute to join the mission, which meant flying from South Korea to join Archbishop Tutu in Brussels and then flying on together to Abidjan to meet up with Kofi Annan.

When we arrived in Abidjan, the security situation was tense; we were to be protected by UN peacekeepers. The UN officials briefed Kofi about a rumour going around that the former president was being ill-treated in detention and that this was adding to the tension. Kofi ascertained where Gbagbo was being held, several hundred kilometres to the north, and asked the United Nations to provide transport for us to go and see him. We found him under house arrest in a luxurious home. Far from being tortured, he was being well treated and was in good health, although he seemed to underestimate his own responsibility for the violations of human rights that had occurred following his refusal to accept the election result.

On our return to Abidjan, we were able to quell the rumours of Gbagbo's ill-treatment and this defused the situation somewhat. We subsequently had a meeting with President Ouattara and his full cabinet in which we raised, among other issues, the sensitive matter that forces supporting Ouattara were also alleged to have committed serious human rights violations. We made the point that fair processes and the rule of law would have to be applied even-handedly; it couldn't be just a victor's justice. We met separately with opposition leaders, who were effectively under house arrest, and heard their concerns as they told us in no uncertain terms that they feared they would be indicted and others would not. We later emphasised this point about the imperative to avoid victor's justice, to former prime minister Charles Konan Banny, who, while we were there, was nominated to chair the truth and reconciliation commission being established. The Elders continued to monitor developments in Côte d'Ivoire for fairness.*

As well as concentrating on particular countries or regions – including Côte d'Ivoire, South Sudan, and the Middle East – the Elders focused on broad themes to highlight values we believed were vital to a fair and peaceful world. At our initial planning meeting in May 2007 we had determined that the Elders would be guided by the Universal Declaration of Human Rights and by our own vision statement (and we would be supportive of the UN Millennium Development Goals). To mark the sixtieth anniversary of the Universal Declaration, the Elders embraced a campaign, which was developed by a group of supporting organisations, called 'Every Human Has Rights'. The campaign, whose purpose was to explain the different human rights and increase consciousness of them, was launched in Cape Town by Arch, Graça Machel, and me on 10 December 2007. We organised events every month during 2008, which were undertaken by large bodies such as Amnesty International and Save the Children, highlighting different human rights at each. The year ended with an event in Paris, where two powerful speeches were given, one by Jimmy Carter, recalling with compelling honesty how he fought racism in his native state of Georgia, and the other by Stéphane Hessel, French Resistance hero and concentration camp survivor, who later, as a diplomat, had been heavily involved in the drafting process of the Universal Declaration. He gave some indication of how he was going to stir the conscience

---

* Gbagbo was subsequently indicted before the International Criminal Court and transferred for trial to The Hague in November 2011.

of a whole new generation with his manifesto, 'Time for Outrage: *Indignez-vous!*'

Civicus, an umbrella grouping of civil society organisations based in South Africa, subsequently took on the 'Every Human Has Rights' logo and materials to use for education about, and to raise awareness of, human rights issues.

Another broad theme the Elders adopted was to tackle discrimination against women and girls, so pervasive in all parts of the world. Early on we discussed the underlying factors in depth, and concluded that, in different ways, religion or tradition was often distorted or misused to subjugate women and put them in inferior positions. Graça Machel spoke eloquently about the way in which women and girls are treated as 'second-class citizens' in so many countries, explaining that this attitude is reinforced by harmful traditional practices such as child marriage and female genital cutting, and that these traditions go deep, so that mothers who have endured these practices nonetheless inflict them on their daughters. I volunteered some of my own experience in dealing with the Catholic Church in Ireland over the years on how its traditional practices had undermined equality for women. If anything, Jimmy Carter's was the strongest voice of all in condemning religious and traditional practices that discriminate against women. He had seen this first-hand in decades of work in developing countries, and he and his wife, Rosalynn, had walked the walk by challenging inequality in the Baptist church to which they belonged in Georgia, eventually leaving it to join a more progressive church. We decided to issue a strong statement calling on religious leaders and traditional chiefs to advocate for and support the full equality of women and girls.

Our CEO, Mabel van Oranje, identified the practice of child marriage as an issue the Elders could raise awareness of and advocate against. When the Elders team subsequently reported its research on the prevalence of child marriage, we were all taken aback by just how serious a problem it is in the twenty-first century. Some ten million girls are married every year, long before they are ready physically. Lack of power in the marriage means that these girls cannot abstain from sex or insist on condom use. This results in serious health risks such as obstetric fistulas from pregnancy before the girl's body is fully developed, and the contracting of sexually transmitted diseases, including HIV. It also means girls leaving school at a young age to the detriment of their education and childhood.

The practice is poverty-driven: once a family can marry off its daughter, she is out of the family; often, the older the girl is, the larger the dowry expected; and there is also a question of 'honour', to avoid any relationship between an unmarried girl and a member of the opposite sex, which would bring shame upon a family. Also, and perhaps most challengingly, child marriage it is a deep-rooted custom or tradition within communities: even when parents under-stand that it is harmful for their daughters, it is hard for them to break with tradition. The link to poverty ties in with my own think-ing on development and rights: poverty-reduction efforts are required by governments to reduce and prevent the practice.

The Elders agreed upon a strategy whereby our team would be in touch with civil society organisations in different regions, and with bodies such as UNICEF, already working on the issue, with the aim of forging a new global partnership to tackle child marriage. We set up a regional meeting in Ethiopia of organisations working on child marriage in Africa and South Asia, and we committed a group of Elders to visit the Amhara region of Ethiopia, where there was a high prevalence of early marriage. Arch, Grö Brundtland, and I took part in this visit, and Graça Machel joined us for the subsequent regional meeting in Addis Ababa. The Ethiopian minister for health, Tedros Ghebreyesus, informed us that Ethiopia had a good law on marriage – marriage was prohibited for those under eighteen: the difficulty was in enforcing this law.

When we travelled to the Amhara region, we discovered, talking to the regional commissioner, a strong woman committed to chang-ing the practice, that here the average age of marriage for girls was twelve. She spoke about how difficult it was to change this type of custom or traditional practice, and how it could be done only if the whole village were involved – the Imams, the men who were elders, school boards and teachers, the girls themselves and their families. The whole village also needed to embrace the social benefits: if girls stay in school they will be more productive, contribute to society through work, and make better mothers when educated and more mature.

In one village we were able to talk to some couples. I met with a sixteen-year-old girl, Anama, and her husband of a year, who looked to be in his late twenties. I asked her gently to tell me about her wedding day. She looked at me with the saddest eyes and said, simply, 'I had to leave school.' It brought home to me the reality for so many girls: One day she had been a happy schoolgirl, learning, playing with her friends. The next day, she was told by her parents that tomorrow she would be marrying a man she had never met, her family would pay a dowry, and she would live in that man's household and do whatever was expected of her.

One of our Elders team, a young woman, Françoise, had chatted to some of these girls on an advance visit and told them that she had just become engaged herself. They looked at her, surprised. 'What age are you, miss?' 'Nearly thirty,' Françoise replied. They couldn't believe it.

Following our visit to the Amhara region, we returned to Addis Ababa, where organisations working on the issue of child marriage from African and South Asian countries met and agreed to work together, launching a global partnership that was later called 'Girls Not Brides', to highlight the issue of child marriage and campaign for its prevention. A number of philanthropic organisations committed to providing funding, and the Elders team recruited new staff who would eventually take it forward as a separate entity.

As part of consolidating that global partnership, an Elders team consisting of Arch, Grö, Ela Bhatt, and me, made a visit in early 2012 to the Bihar region in India where early child marriage is also prevalent. Through Ela, hosting us in her own country, we were able to learn more deeply about the complexity of the problem. We were encouraged that government representatives both in Delhi and in the state of Bihar agreed to meet us and engage in discussions about child marriage.

The federal law in India provides that girls cannot marry before eighteen and boys before twenty-one, but again, enforcement is difficult in various parts of India. As Ela emphasised, we were there to listen to girls affected by child marriage, their parents, their teachers and community leaders – and amplify their needs and concerns in our conversations with government, media, and influential people.

I remember, on a visit to a rural area, the bright beautiful faces of girls of thirteen and fourteen as they told us of how they were negotiating with their parents to be able to stay in school another year and not be married until at least fifteen. Still a depressing prospect, but at least, in this school, they were being supported in their defiance of their parents, trying to push the marriage age out as far as they could. If one girl had a problem with her parents, they told us, the rest of them would go along and support her and plead that she be allowed to stay on in school with them.

As 'Girls Not Brides', now involving over 120 organisations, became independent of the Elders, we moved on to discuss other ways to tackle discrimination against women and girls, with a particular focus on the role of men as traditional and religious leaders.

More and more during these years of working with Realizing Rights and with the Elders, I was becoming aware of the connection between climate and poverty. As honorary president of Oxfam, I was invited to sit on a 'jury' with Archbishop Tutu in November 2007 to listen to the accounts of five African farmers Oxfam had brought together. The exent took place in Cape Town. Of the five farmers, four were women – a reflection of the fact that 80 per cent of farming in Africa is done by women. I remember the essence of each of their stories: The first was a rooibos tea farmer from South Africa, accompanied by her husband. She had made quite a good living until floods destroyed her tea crop, as the weather pattern in the region changed. The male farmer was from Kenya, a pastoralist who had seen his two hundred goats reduced to twenty because of drought.

I knew the next witness, Constance Okellet from northern Uganda, who was later to become one of the 'climate wise-women' bearing public witness at the Copenhagen climate conference and elsewhere. I have often quoted her story, as she tells it so well, of growing up in a village in northern Uganda when they had regular seasons, sowed and harvested crops, and had adequate food. But about six or seven years earlier, in 2000, everything had changed. There were no seasons, only long periods of drought and flash flooding (which destroyed the school). Constance, a mother of eight, along with other women in the village, formed a group to try to hold their community together. 'We thought God was punishing us,' she said in her slow, dignified voice, 'but then I learned through Oxfam that it was the lives of rich people causing this.' The fourth witness was a woman from Mali whose cotton production had failed because of drought, and the last was a woman from Malawi whose evidence was the most desperate of all: 'Flooding has destroyed our crops and our village. Some of the women are selling themselves, but I cannot do that as I am HIV positive.'

Archbishop Tutu had greeted this group of farmers in several local languages, to their pleasure, as we were photographed with them before we began. Now he sat slumped in his chair looking dejected by their accounts of how their lives had been devastated. So I decided to challenge these five witnesses a bit. I told them I had grown up in a poor part of Ireland where there were small farmers, and that in my experience, farmers always complained about the weather. Was that what they were doing, complaining about a bad patch of weather? I smiled to show that my question was not a hostile one. Their response, spoken by Constance with great dignity, stayed with me. 'This is different,' she said. 'This is outside our experience.'

What is 'outside our experience' in an African village? I estimated it might be beyond two hundred years of word passed down from grandmother to grandchild, through the generations.

When, in July 2011, at the request of Irish aid agencies, I revisited Somalia, I couldn't help thinking that on every count it was worse than when I had been there as president of Ireland in October 1992. Then, when I visited in October, it was the end of the dry season, and the rains were expected. This time it was July, and while we were there the United Nations declared famine in two regions of the country. We were informed that twenty-eight thousand children at least had already died of hunger in the twenty-first century! And many of those who survive are stunted and suffer permanent effects in terms of physical and intellectual development. When I was there in 1992 it was possible to talk to the warlords and seek to impress upon them the need to let the food into the feeding stations. It is not possible to talk to Al-Shabaab, the Islamist militant organisation linked to al-Qaeda; its members are vicious with their own people and hostile to foreigners. Climate change was an issue that did not come up at all when I was there in 1992. But in July 2011 it was clearly impacting the Horn of Africa, which had had its eight hottest years in succession since records began. What was going through my mind was: nineteen years since the last, but it will be only four or five years before the Horn of Africa will have another incredibly severe drought. And it is not just the population of Somalia that suffers. As I write, it is estimated that there are about thirteen million people in the Horn of Africa at risk.

The various strands of my involvement in working with African countries, whether on health issues, decent work, corporate responsibility, or women, peace, and security, led me to an understanding that there was a constant negative factor impeding the development of poor countries: the negative impact of climate change. I saw first-hand how devastating long periods of drought followed by flash flooding – that is, the absence of predictable seasons – can be to poor communities. It undermines their food security, their health, their schools.

At a conference on climate change held in Rwanda in 2007, I learned that the whole African continent was responsible for less than 4 per cent of greenhouse gas emissions, yet I knew from my own experiences that the impact was much more serious than in richer countries. I was increasingly disturbed by what I describe as the 'Ah, but . . .' sentences that I heard over and over again: 'Ah, but things are so much worse. We have no seasons anymore. We have prolonged drought and then flash flooding', or 'Ah, but we used to be able to predict the rainy seasons, and sow our seeds, but not anymore . . .' These communities are not responsible for the emissions causing climate change, and yet they are disproportionately affected because of their already vulnerable geographic locations and their lack of climate resilience. This, too, is an issue of justice, and needs to be addressed.

Looked at through the lens of justice, there is another challenge. It isn't just governments that are causing greenhouse gas emissions, it is us, the people: through the way we do business, our transport, the way we heat buildings, what we choose to eat or wear, habits in the family regarding whether waste is recycled or switches turned off when not in use. We have to change our lifestyles, wean ourselves off bad habits of consumption.

Recently, in preparation for the world conference on sustainable development in June 2012, to mark twenty years since the first Rio Earth Summit, four Elders (Arch, Grö, Fernando Henrique, and I), engaged in a public debate with four young people, from Brazil, China, Nigeria, and Sweden, and found ourselves influenced by the passion and sense of urgency these young people brought to the debate. It is their future, and they know we are not on course to stay below the two degrees above pre-industrial levels of warming that scientists have warned is necessary for a safe world, and that governments agreed to at the climate conference in Cancun in December 2010.

This public dialogue about climate and the environment is challenging both old and young to think deeply about how we will forge an equitable sustainable future, one that safeguards our planet and is fair to a population predicted to rise to nine billion or more by 2050, the increase coming mainly from those large countries with emerging economies such as China, Brazil, and Nigeria. As well as the now-iconic image of a polar bear drifting on a melting icefloe, we need the face of a poor subsistence farmer or indigenous woman, a face full of worry because her livelihood and way of life are being destroyed by climate change.

I have been on my own journey in pondering these issues; a journey that has taken me back to live in Ireland. It had always been my intention to come home when my work with Realizing Rights was completed. Now I had a new cause to pursue.

19

Connecting the Global and the Local

Our experience is not just our own property – it must be shared.

Ela Bhatt

I TRULY loved living and working in New York, and it was with sadness that I bade farewell to my colleagues. But I was eager to return to Ireland, where I could develop my ideas around the connections between human rights and climate change. What I wanted to do when I got home was clear, as if everything I had learned was coming together to make whatever contribution I could to tackling the greatest potential threat, not just to human rights but to the future of the planet itself.

The Ireland I returned to in December 2010 was struggling to recover from financial collapse and the subsequent bailout by the European Union, European Central Bank and International Monetary Fund. The downward spiral of the economy happened quickly; people are angry, and have every right to be. At the time of writing, the country is on its knees. People are hurting: many households now find themselves in negative equity, unable to pay their mortgages and stunned by the sharp fall in house prices. Small businesses have closed and many are still closing, and the cutbacks required to meet the financial targets imposed on Ireland as part of the bailout package are hurting the poorest and most vulnerable in particular. Unemployment has once again become a scourge, and emigration is back as a reality for a new generation.

We need to be able to lift our heads again, to recover from the humiliation of loss of economic sovereignty. As a former president, it would not be appropriate for me to engage in any political or public role on our own financial woes, but I felt I could play a small part in an area where Ireland has great global credibility: in tackling hunger and supporting the poorest developing countries. Despite the hardship Irish people themselves are enduring – job insecurity, difficulty in meeting household bills – aid agencies such as Concern, Trocaire, and Oxfam Ireland continue to be supported in their work, as I saw when they invited me to join them in Somalia in July 2011, to help draw attention to the looming famine there.

So, having made the decision to focus on climate justice when I returned to Ireland, what would be the best way to do this? I had experience of sitting on the board of two foundations that were not grant making, but that sought to achieve specific objectives: the Mo Ibrahim Foundation, which supports good governance and leadership in Africa; and Silatech, established by Sheika Moza bint Nasser of Qatar, which addresses youth unemployment in the Middle East and North Africa. There was no tradition in Ireland of former presidents establishing foundations, but I decided I would establish a charitable foundation on climate justice for which I would work pro bono. I consulted Mo Ibrahim, whom I admired not just for the impact his foundation was having in African countries, but also for his directness and occasional political incorrectness. His forthright views on tackling corruption and having accountable leadership in African countries – as an African from Sudan – were so refreshing and effective.

Mo asked me what the foundation would be called, and when I hesitated to use my own name, his response was emphatic. 'Mary, do you recall the public relations budget I needed to promote my foundation? If you call it the "Mary Robinson Foundation – Climate Justice", you won't need to spend a penny to promote it – everyone will know it has something to do with human rights and climate change.'

The concept of climate justice excited me because it married the principles of human rights with issues of sustainable development and responsibility for climate change. I believed it was possible to adopt a holistic approach to tackling this problem: poor people have a right to develop, but if the millions entering the middle class in China, India, and African countries adopt our fossil-fuel-dependent lifestyle, then it will be impossible to keep global warming within the safe levels defined at Cancun in 2010. We are all in this together: not only is there a moral obligation on rich countries to reduce urgently their dependence on fossil fuels such as oil and coal, but it is also in the interest of global survival that rich countries share renewable energy technology with poor countries, and support them in adapting to become more resilient to weather shocks.

I had come to climate justice as a human rights advocate. Now, in my mid-sixties, I was starting a steep learning curve on the science, politics, and economics of climate change.

The establishment of the foundation as a charitable entity and the recruitment of the initial board members would have been impossible without the enormous commitment of Bride Rosney, who once again gave me invaluable support. The foundation was established as part of the alliance of innovation between Trinity College and University College Dublin, and both universities have two nominees on our board. Given my long association with Trinity, and my continuing role as chancellor of the university, I had sought advice and support from the then provost, John Hegarty, and it was agreed that the foundation could be located on the ground floor of a Georgian house in South Leinster Street that Trinity had acquired. We received funds from philanthropic organisations, initially for establishment costs and then for programme and ongoing operational costs.

We constantly interrogate ourselves as to whether we are truly looking at climate change from the perspective of the poorest, bringing out the justice elements and seeking solutions based on equity and the right to development. Development is not possible without energy, but in the twenty-first century it has to be clean, affordable, and sustainable energy. The stark reality is that approximately 1.3 billion people – out of a world population of 7.0 billion – have no access to electricity and rely on kerosene and candles for lighting. And 2.7 billion people (mainly women) continue to rely on firewood, coal, or animal dung for cooking and heating, with grim consequences for their health as they, and the children near them, inhale smoke into their lungs.

In a world of such shocking inequality, how could a tiny foundation based in Ireland make a difference?

Drawing on existing international documents such as the Universal Declaration of Human Rights and the UN Framework

Convention on Climate Change, we brought together an expert group to help us draft climate justice principles to guide our work.

We recognised that provision of new types of off-grid energy – solar lights, clean cookstoves, small hydro projects – was becoming a crowded field, with social entrepreneurs and even large enterprises finding ways to sell to the so-called bottom or base of the economic pyramid, those with the least ability to pay. It seemed to us that they were focusing on the potential market of five hundred million or so who had mobile phones but no electricity in their homes. But what, we asked ourselves, about the further approximately nine hundred million who had neither mobile phones nor electricity in their homes? How could we find ways to reach them, and scale up their access to electricity? We realised that in a number of countries various kinds of government-organised social protection systems provide support for the poorest sector in their society. With the help of funding from philanthropic organisations, we brought officials from social protection systems in India, Brazil, and Mexico together with social entrepreneurs and others who provide access to off-grid energy and have ideas about local micro-credit and financing. This expert group found that there were many potential synergies between social protection and access to energy, and that these could and should be acted on. Now we are monitoring how the participants have changed their way of working, and are looking for means to develop these synergies on energy access in various countries.

Another obvious focus of our work would be to highlight the way climate change exacerbates hunger and malnutrition. Although I try to resist invitations to take on new commitments, I made an exception when asked in April 2012 by UN secretary-general Ban Ki-moon to join a leaders group on food and nutrition security, 'Scaling up Nutrition' (SUN), as the impact of climate change on food security is another issue my colleagues and I are working on. The right to food is one of the most basic rights of humankind, but hunger is still a reality for millions, with many systems of food production simply unsustainable.

We are in an era of rapid population growth. When the United Nations estimated that the world's population had reached seven billion in October 2011, we were reminded that it reached six billion only twelve years before, and we would reach eight billion in less than eighteen years. With the world's population set to march on to nine billion by 2050, agricultural production will need to increase by 70 per cent in order to meet demand. And yet climate change is undermining farming and food production for millions of subsistence farmers already – the great majority of whom are women.

While women in weaker economies – being the farmers, the providers in the home, and the backbone of their communities – are the most affected by the injustices of climate change, women in general have a particular role to play in climate activism. Women are more likely to change behaviour – in the family, in the home, in children going to school, in the local community – through a changing of choices. An example is the '1 Million Women' initiative in Australia, whose goal is to take one million tonnes of carbon out of the atmosphere through measures adopted by individual women in their homes and workplaces.

Women, as part of nurturing, tend instinctively to think intergenerationally. This was brought home to me on a very personal level when I became a grandmother. When I am occasionally disheartened by the scale of the problems we confront, I draw upon the sense of that complete shift in my own time horizon when Rory was born. Ever since, I have been thinking a hundred years hence, and it fires me up for renewed work.

There is progress everywhere. Women at the grass-roots level are not only organising more but making their presence felt. A good example is the Global Alliance of Waste Pickers: these are the poorest of the poor, many of whom are women, who recycle the waste they pick from dumps to make a tiny living. Through e-mail and social media, they have formed a global organisation, supported by a group called Women in Informal Employment: Globalizing and Organizing (WIEGO), run out of Harvard University.*

Further significant breakthroughs will come as a result of the constant linking of those women who have access to political power

---

* The Mary Robinson Foundation has developed a network of women working at the grass-roots level, and Waste Pickers has joined that.

and law-making – who are at the table, if you like, nationally and internationally – with women on the ground, who know the relentless problems in a very direct way and who are now beginning to feed in and share that experience, gathering strength in numbers even if their status in their communities is quite low. As women leaders at different levels, the challenge for all of us is to harness those strengths more effectively in order to alter the narrative and priorities on climate change. Women need to emphasise the urgency of addressing *now* trends that could become unstoppable and lead to catastrophic consequences for millions. As mothers and grandmothers, we need to shake political leaders out of their short-term thinking, and insist that they, too, think intergenerationally for the sake of the future of the planet.

The issue of population growth is often raised in discussions about climate change. Frequently it is in the form of a question: what are we going to do about population control? But the words *'population control'* are highly objectionable to people and governments in developing countries, evoking mistakes of the past on family planning measures and implying that steps need to be taken from the outside to curb population growth. So I answer by pointing out the need for sensitivity on this issue, and the need to support what has been shown to work in reducing family size: educating girls and women, including on their reproductive health, and ensuring that health systems are functioning to reduce maternal and child mortality. If girls and women are educated, and are confident that their babies will survive the first five years and beyond, they are much more likely to have fewer children.

Access to education on reproductive health and the means of family planning are basic rights of every woman.* However, it is estimated that more than two hundred million people (mainly women) have unmet needs for family planning products. There are, of course, strong ideological views on this issue in different parts of the world, particularly if linked to abortion, but family planning

---

* As they were so proclaimed at the Cairo International Conference on Population in 1994.

itself should be supported in the twenty-first century. Having promoted family planning legislation in Ireland as long ago as 1970, it troubles me that this is a fight that still has to be fought globally – it is one on which I am prepared to take a stand.

On the issue of food security, having seen many starving children on countless heart-rending occasions, it is not the grim statistics that interest me, but how we can get our priorities right. I am determined to bring out both the injustice of the impact of climate change, and the fact that, in our twenty-first-century world, we have everything we need to ensure that no family has children waking up hungry every day, becoming stunted so that they will never reach their physical and intellectual potential. I believe we will succeed only if we adopt a participatory, bottom-up, climate justice approach that brings out the human rights and gender equality aspects, with particular focus on the economic empowerment of women. If women are empowered to earn an income, or benefit from a minimum protective financial safety net if not in a position to earn, children will not go hungry.

In December 2010, together with colleagues and wearing both my Realizing Rights hat and my new hat as president of the Mary Robinson Foundation, I worked with the government of Mexico to organise an event for women leaders on gender and climate change at the climate conference in Cancun. The idea emerged of building on the troika of women, Connie Hedegaard, Patricia Espinosa and Maite Nkoana-Mashabane, who had presided over three climate conferences in a row.

The elements for building an innovative alliance of women leaders on climate change were already there. During 2011 the 'troika' expanded to a 'troika plus' of women leaders on gender and climate change; there were more than fifty members at the climate conference in Durban at the end of the year, including some supportive male ministers.

In New York in March 2012, during the Commission on the Status of Women, Michelle Bachelet, executive director of UN Women, and other troika plus members listened to grass-roots rural women describe their efforts to secure affordable renewable energy for their homes and work: how expensive it was, how difficult, but how it would transform their lives. The troika plus members promised to bring their concerns to the Rio+20 conference on sustainable development in June 2012.

The mood in the room during our exchanges had been charged with energy and purpose. The fact that rural women from networks in Sudan, Ghana, Nepal, and Bangladesh were being listened to for several hours encouraged them to speak freely about their frustrations and their determination to gain access to clean cookstoves, solar lights, and other equipment that would change their lives. My colleagues and I are determined that the troika plus of women leaders on climate change and the networks of women at the grassroots level continue to interact in a way that impacts all aspects of how climate change is addressed.

Working from Ireland adds to my sense of excitement and pride in what we are trying to accomplish. The location of the foundation, based in Trinity College, has another benefit. The foundation is helping to prepare a module on climate justice as part of a new masters degree in development practice. The degree is being put together jointly by Trinity and UCD, as part of a wider network that takes in more than twenty universities worldwide. I enjoy giving occasional lectures as part of this programme. It offers me a chance to get back to teaching, which has been one of the great pleasures of my professional life. Modern technology is enabling universities to connect and share knowledge in new ways, which can generate the understanding and empathy needed to enable the younger generation to avert the looming climate change catastrophe for which mine, and prior generations are responsible.

While I can observe the harsh reality that so many Irish families face, and the inequity in how the burden is shared, I continue to honour the tradition that, as a former president, I should remain outside the cut and thrust of politics. As an observer, my main reference points are once again local: back in my beloved County Mayo. I see the hardy resilience people there are showing, the spirit of *meitheal* coming alive again, those acts of neighbourly kindness to try to fill the gaps in services brought about by cutbacks. Tough times bring out the character of a people. The impact of the economic downturn on Ballina is typical of that in towns and villages throughout Ireland in recent years. Many more families need the free meals provided, and the second-hand clothes. There is a rising demand for training in different skills, literacy courses, and cookery classes.

In September 2011, I was invited to a meeting in Ballina of service providers in the town and wider area. About eighty people crowded into the room and conveyed the struggles they were having, to cope with both budget and staff cutbacks. In some cases, retired teachers were roped in to volunteer; in others, efforts were made to minimise the negative impacts on those with disabilities, families with special needs children, the elderly living alone in remote areas, traveller families – the list went on. Those who spoke didn't politicise their accounts – although, in a number of cases, I could see the anger barely being kept in check. Each of them was doing human rights work, although they might not have put it that way. What came through to me was the enormous respect and affection – even love – they had for those who were suffering from the cutbacks, and their determination to try to minimise that suffering. All these people seemed so young to me, so that I really felt like an elder.

Looking around at their bright, committed faces I felt a sense of pride in their resilience – a source of hope for the country. Some were employed in state or local authority schemes, some by local organisations; some were volunteers. None of them thought he or she was doing anything special or out of the ordinary. But the way they were coping brought home to me vividly how everyone can make a difference. An important factor, which was clearly present in this group of people, was empathy with those they served. Empathy leads to respect for, and a visible understanding of, the situation of the other – an affirmation of their dignity and worth. That was the example my father gave me all those years ago, when he stood on the doorstep with the wife of a patient, bending his head towards her and listening to her worries and concerns.

★   ★   ★

For me, being back in Ireland, particularly in the West of Ireland, means being immersed again in our history, in memory, the link with the past. That honouring of the past can also be part of the future, connecting with the Diaspora and beyond. I witnessed this recently in our local village.

Easter Sunday, 8 April 2012, began early. My brother Adrian collected me from my home at 6.00 a.m. and we drove to Addergoole Cemetery, beautifully located on the shore of Lough Conn, and well maintained by the local parish. Cars were arriving from all the small roads leading to the graveyard, and it was clear that this first Easter dawn Mass in Addergoole was a popular idea. Addergoole Abbey, possibly dating back as far as the sixth century, founded by Saint Kieran, is an ivy-covered ruin, with three walls standing and the back wall largely collapsed, allowing for a wonderful view out onto the lake. There are a number of flat gravestones on the ground, and on these the gathering crowd stood as we assembled for Mass, with as many again outside, listening to the familiar lessons and gospel. As dawn broke, it was an unforgettable moment, the only downside being a host of midges, which rose in clouds from the ivy and attacked the warm human bodies packed close together.

Later that morning, I was invited by the Addergoole Titanic Society to help our local village of Lahardane launch a cultural week to commemorate the fact that, in 1912, no fewer than fourteen villagers sailed on the *Titanic*, eleven of them perishing when it sank. The first event was to visit a memorial park that Taoiseach Enda Kenny, himself a Mayo man, would formally open on the centenary date, 15 April. The local people understood symbolism. There was a light in a specially constructed gable window that I was asked to switch on, acknowledging the light in the window of Áras an Uachtaráin to connect with the Irish Diaspora.

A large crowd had gathered, of local villagers and those who had come specially from the United States as relatives of the 'Addergoole fourteen', the eleven women and three men who had travelled to Cobh in County Cork in April 1912 to board the *Titanic*. As Nick and I walked from the memorial park down to the church, blessing our luck that although cloudy it wasn't raining, we saw an even bigger crowd waiting. I had been introduced to the fourteen young people from the village who, dressed in 1912-style clothing, would re-enact the journey of hope that ended so disastrously. After short speeches and a blessing from the parish priest, the fourteen climbed into carts drawn by fine dray horses and made off towards Castlebar while the people of Lahardane and visitors waved them goodbye. A moment of silence and sadness accompanied these young people as they rode away to enact the pageant.

For me, there was a deep sense of being home, both physically and spiritually. Nick and I have a plot in Addergoole Cemetery. We are in no hurry, but that is where we hope to end up. For all my travels around the world, I am a very rooted person. There is a special joy in being recognised first and foremost as my father's daughter, or having memories of my mother or grandparents. The privilege of belonging, and knowing that you have come full cycle in life, brings an inner peace of mind.

Acknowledgements

I AM conscious that there have been distinct stages in my life's course and that at each stage I have stood on the shoulders of many others. I am deeply grateful to all of those who have helped shape my life: parents and grandparents, brothers, wider family and friends, colleagues who worked with me during each of these stages and mentors.

I would like to thank Scott Jerbi, Ronan Murphy and Bride Rosney for all their time and effort helping me with this memoir, and also Eavan Boland, Inez McCormack and John Horgan.

Thanks also to Mary Baylis, Cecilia Canessa, Celine Clarke, Hillary Johnson, Michael Keohane and Anne Quinlan.

We consulted a number of books on the subject and wish to acknowledge their authors: Olivia O'Leary and Helen Burke, the late Kevin Boyle, Fergus Finlay, John Horgan, Michael O'Sullivan, Emily O'Reilly and Lorna Siggins. I would also like to acknowledge the library at Trinity College, Dublin.

I am grateful to Judith Rodin and the Rockefeller Foundation who encouraged me in a very practical way by offering a residency in Bellagio.

I would like to thank my agent Lynn Franklin and her UK colleague Mary Clemmey for all their advice and encouragement, my supportive editors Rowena Webb, Helen Coyle and the team at Hodder & Stoughton, and in the United States, George Gibson.

This was, in a very real sense, a family enterprise. I would like to thank my brothers Henry and Adrian for their help and support with this memoir, and for sharing memories of our childhood. Thank you to my son Aubrey for his technical help with the photographs and illustrations. In particular, I would like to thank my husband and great ally Nick who read each draft with a beady eye, applied his red pencil liberally and helped choose the illustrations to accompany the text.

Above all, I want to thank my daughter Tessa for taking on this enterprise with me – it has been a pleasure to work so closely together over many months. And a special thank you to Robert, Rory and Amy for their patience and support.

My warm thanks to all those who helped with this memoir, of course I acknowledge that any mistakes are my own.

Picture Acknowledgements

Family Archives: 1, 2, 3/above right Peter M. Robinson, 4/Nick Robinson, 6 above, 7 above/Nick Robinson.

Additional Sources
Thierry Boccon-Gibod/The Elders: 15 above left. Photothèque de La Commission Des Communautés Européennes, Bruxelles: 5 above. Corbis: 14 centre right/Juda Ngwenya. Getty Images: 16 above right/Attila Kisbenedek. The Irish Times: 6 above right/Jack McManus, 8 below, 9 above/Pat Langan, 10 above/Frank Miller, 11 above/Eric Luke. Jennifer McCarthy: 15 below. Jeff Moore/The Elders: 16 below. L'Osservatore Romano/Servizio Fotografico: 5 below, 12 below. Press Association Images: 10 below/Martin Keene, 14 below right/Johnny Green, 15 above right. Reuters: 13 above/Pascal Volery, 14 above. Rex Features: 16 above left. Mick Slevin: 9 below. Derek Speirs: 11 below. Courtesy of the White House: 12 above, 14 below left. Henry Wills/Western People: 7 below, 8 above.

Every reasonable effort has been made to trace the copyright holders, but if there are any errors or omissions, Hodder & Stoughton will be pleased to insert the appropriate acknowledgement in any subsequent printings or editions.

Cartoons within text
Nick Robinson: 83. Scratch (Aongus Collins): 167. Martyn Turner/The Irish Times: 162, 178.

Text Acknowledgements

Extract taken from the speech 'In Our Hands' by Eleanor Roosevelt was taken from the 1958 speech delivered on the tenth anniversary of the Universal Declaration of Human Rights.

Extract from 'The Second Sex' by Simone de Beauvoir was published by Random House.

Extract taken from a speech by Albert Einstein.

Martin Luther King quote is reprinted by arrangement with The Heirs to the Estate of Martin Luther King Jr., c/o Writers House as agent for the proprietor New York, NY. Copyright 1963 Dr. Martin Luther King Jr; copyright renewed 1991 Coretta Scott King.

William Gladstone extract was taken from a speech, 10 April 1888.

Extract taken from the poem 'Domestic Interior' by Eavan Boland from Outside History: Selected Poems 1980 – 1990. Published by WW Norton.

Extract taken from the poem 'Asphodel, That Greeny Flower' from Selected Poems by William Carlos Williams. Published by Penguin Books UK.

Extract taken from the poem 'I am of Ireland' by W.B. Yeats (1865 – 1939)

Extract taken from the poem 'Psalm' from Selected Poems by Paul Celan, translated from the German by Michael Hamburger. Published by Penguin Books UK.

Extract taken from the poem 'The Cure At Troy' from Open Ground: Selected Poems 1966 – 1996 by Seamus Heaney. Published by Faber and Faber.

ACKNOWLEDGEMENTS

Extract taken from 'Three Guineas' by Virginia Woolf (1882 - 1941).

Extract from 'Markings' by Dag Hammarskjold, translated by W.H. Auden, Leif Sjoberg. Published by Faber and Faber.

Extract taken from the poem 'A Disused Shed in Co. Wexford' from Collected Poems by Derek Mahon. Published by The Gallery Press.

Extract from "The Drowned and the Saved" by Primo Levi. Published by Penguin Books.

Extract taken from the poem 'From the Republic of Conscience' from Open Ground: Selected Poems 1966 – 1996 by Seamus Heaney. Published by Faber and Faber

Extract quote by Dorothy Q. Thomas, taken from 'The Unfinished Revolution', edited by Minky Worden. Published by Policy Press.

Extract taken from a speech by Nelson Mandela.

Extract taken from the essay 'Culture and religion in Ireland, 1960 – 2010' by Professor Enda McDonagh.

Extract taken from the poem 'A Disused Shed in Co. Wexford' from Collected Poems by Derek Mahon. Published by The Gallery Press.

Index